WILSON'S CHINA

A CENTURY ON

To Ian Bond, Michael Heathcoat-Amory, Lord Dulverton and Lady Lennox-Boyd –
without your help this book would not have been possible.

WILSON'S CHINA

A CENTURY ON

MARK FLANAGAN & TONY KIRKHAM

Kew Publishing
Royal Botanic Gardens, Kew

PLANTS PEOPLE
POSSIBILITIES

First published in 2009 by
Royal Botanic Gardens, Kew,
Richmond, Surrey, TW9 3AB, UK
www.kew.org

ISBN 978-1-84246-394-9

British Library Cataloguing in Publication Data
A catalogue record for this book is available from the British Library

Production editor: Michelle Payne
Typesetting and page layout: Christine Beard
Design and cover design: Hina Joshi
Publishing, Design and Photography, Royal Botanic Gardens, Kew
Front cover image: © Tony Kirkham

Printed and bound in Italy by Printo Trento

Mixed Sources
Product group from well-managed forests and other controlled sources
www.fsc.org Cert no. CQ-COC-000012
© 1996 Forest Stewardship Council

The paper used in this book contains material sourced from responsibly managed forests, certified in accordance with the FSC (Forestry Stewardship Council)

For information or to purchase all Kew titles please visit www.kewbooks.com or email publishing@kew.org

Kew's mission is to inspire and deliver science-based plant conservation worldwide, enhancing the quality of life.

All proceeds go to support Kew's work in saving the world's plants for life.

CONTENTS

Acknowledgments   7

Foreword   9

Map   12

Introduction   13

Chapter One: Ernest Henry   27

Chapter Two: Wawu – Luding, Dadu – Kangding   33

Chapter Three: Mystery towers of Danba   67

Chapter Four: The dreaded Pan-lan Shan   99

Chapter Five: Dig the bed deep, keep the banks low   129

Chapter Six: A middle way to Songpan   143

Chapter Seven: Nemesis in the Min Valley   185

Chapter Eight: Saving Wilson's plants   213

Epilogue   229

References   247

Selected bibliography   251

Index   253

ACKNOWLEDGEMENTS

A book involving three countries on three continents is always likely to rely on a great deal of help from far-flung friends and this book is no exception. Without the help of generous people in England, the United States and, most importantly, China, this work would never have even been started.

Firstly, we would like to thank our sponsors Ian Bond, Michael Heathcoat-Amory, Lord Dulverton and Lady Lennox-Boyd. Without their support our ambition to find E. H. Wilson would not have been possible. Not only did their financial support enable this but their interest and enthusiasm kept the project moving forward at times when it seemed it might falter. Thanks also to Roy Lancaster not just for his insightful Foreword but for being a Wilson devotee of long-standing and, therefore, providing us with encouragement, advice and information throughout the time we have been involved with this project. Roy's infectious passion for plants, people and places inspires all those who know him and his generosity with his time and knowledge has won horticulture many new enthusiasts. At Kew, Marilyn Ward of the Library Services was of enormous value in helping to find references and often obscure archive material, which she did with a ready smile. Gina Fullerlove (Head of Publishing) and John Harris were instrumental in bringing this publication to fruition. Thanks also to Michelle Payne, our editor, to Hina Joshi for the design concept and to Christine Beard for the typesetting and page layout. We are grateful to Pat Davies and John Stone for producing the beautiful, detailed route map, which adds so much to the book. A long-standing friend and former colleague from Kew, Chris Grey-Wilson, has patiently assisted us with the identification of images taken in the field at a time when he was already very busy on a Chinese plant book of his own.

In Boston, we are grateful for the competent help of Sheila Connor and Lisa Pearson, respectively Archivist and Library Assistant at the Arnold Arboretum. As custodians of the vast majority of Wilson's professional and personal papers_this branch of Harvard University performs horticulture and botany a great service; Wilson's written legacy is in expert hands. Peter Del Tredici, the Senior Research Scientist at the Arnold, is an old friend who was of inestimable value during our time researching the Wilson archive, he also took us on an interesting visit to Holm Lea, the former home of Charles Sargent, now an exclusive private residential estate.

In China, our thanks are extended to our collaborators and friends without whom the many journeys we have made, and particularly our 'Wilson trip', would not have been possible. Yin Kaipu, Zhong Shengxian and Wang Hangming, along with Mr Liu, planned,

Davidia involucrata. First discovered by the French missionary and naturalist Père David in 1869, but introduced to western gardens by Wilson in 1904. This plant was the catalyst for Wilson's exploration in China in 1899.

coordinated and executed these trips with infinite care and patience, and in the case of the visit to follow Wilson a good deal of on-the-ground research and preparation. In a nice piece of symmetry we can say, with the same degree of warmth and candour as Wilson, that our dealings with our Chinese companions were of a uniformly agreeable nature, 'adding much to the pleasure and profit' of our trips. Zhong Shengxian also maintained his infinite patience in responding to the many email questions and requests for information and clarification that went into the preparation of this book.

Finally, we acknowledge the man who inspired the whole initiative, Ernest Henry Wilson. He has done much to enliven our many journeys to the East and to bring interest and colour into our professional lives as plant collection managers. We hope that with the publication of this book we have repaid some of the debt we owe, given due regard to the great man and opened his legacy to a wider audience.

Mark Flanagan and Tony Kirkham
January 2008

For our return visit to Sichuan in 2008 we again thank our Chinese friends – Yin, Zhong and Wang – for supporting us and looking after us during our time in the earthquake zone.

MF and TK
November 2008

FOREWORD

My first encounter with the world of E. H. Wilson was on the day I left home and walked the two miles through farmland and woodland to Moss Bank Park, on the outskirts of Bolton in Lancashire. It was 1953 and I was about to start on a career in gardening that would take me from being a boy who knew nothing of foreign plants to being a man who would travel the world's wild places in search of them. I was 15 and a little apprehensive as to what might happen but my foreman was determined that I should begin my plant education immediately. With that, we headed for the brick-walled 'Old English Garden' where he proceeded to point out plants of special interest, including a prickly bush he called *Berberis wilsoniae*. 'This comes from China,' my foreman explained, 'introduced to England by a man called Wilson who named it for his wife.' These bare facts meant little to me at the time nor for that matter did the bush, which had yet to display its autumn colour and bore no berries. I felt I ought to respond in some way so I asked the obvious question, 'Who was Wilson?'

In reply, my foreman told me about this young man from a working-class background, 'his father worked for the railways', who rose to become one of the most famous of all plant hunters. 'He made four expeditions to China at the beginning of this century,' my foreman continued, 'and introduced over a thousand new plants to our gardens.' Even to a beginner these facts were impressive and the more I heard of his exploits, and grew or encountered his plants in cultivation, the more I came to realise I had found a role model.

Unlike Wilson, plant hunting did not offer me a living but it did bring a focus to my life, as well as enjoyment, a sense of achievement and a taste for adventure. My greatest ambition was finally realised, on Mount Omei (now Emei Shan) in China's Sichuan Province in October 1980. Wilson had visited this famous mountain with its rich flora of over 3,000 species in October 1903 and I considered the prospect with something akin to awe. To tread where the great man himself had been all those years ago was for me a dream come true and I have never forgotten it. Such personal experiences are not easy to describe adequately but in this book Mark Flanagan and Tony Kirkham have come closer to achieving it than most authors I have read.

Like many another before and since, the authors are fascinated by the life and achievements of this eagle-eyed plant hunter and have followed his travels and shared his adventures through his writings, as well as by their own experiences spent retracing his journeys in the wilds of China. Following in the footsteps of a person you regard as a

Roy Lancaster at the Howick Arboretum with *Betula albosinensis,* one E. H. Wilson's finest introductions.

legend in his field is, perhaps, the ultimate adventure, as Flanagan and Kirkham have discovered through their visits to the country Wilson regarded with some justification as the 'Mother of Gardens'.

A great deal has been written on this most knowledgeable and successful of plant hunters, much of it in diverse publications and fragmentary accounts and recollections. Then there are the several hundred articles, scientific accounts and a dozen or so books authored by Wilson himself. Surprisingly, a full-blown biography including a comprehensive account and assessment of his discoveries and introductions has still to be published. As the authors remark, there are aspects of Wilson's character that even his contemporaries could not fathom. He was very much his own man, a characteristic not unusual in explorers and pioneers. In this book, the authors focus on another aspect of the man that over the years has attracted recurrent discussion: where exactly did his journeys take him? They are, of course, known in broad outline but due to Wilson's use of an anglicised system of naming locations and, perhaps, personal transliterations of native pronunciations, the exact routes of some of his journeys have caused historians more than a few problems.

In seeking to resolve at least some of these questions, Flanagan and Kirkham made use of information in Wilson's field journals, correspondence and, more significantly, his glass plate photographs taken in Sichuan, one of the richest provinces in terms of potential garden plants. These photographs, and I have been fortunate to view the originals at the Arnold Arboretum of Harvard University and at Kew, depict a wealth of subject matter including landscapes, villages, river scenes, bridges, people, plants and, crucially, veteran or exceptional trees. Armed with copies of these photographs and with the benefit of Chinese guides and local knowledge, their efforts to follow Wilson's footsteps, in a jeep through valleys and across mountain passes, on often-rugged terrain, provided an adventure that is well described in this book. In some places, when following the exact route proved impossible, they had to calculate or else make an intelligent guess as to where the route might be continued.

Not surprisingly, given the passage of time and the changes wrought by population pressures, deforestation and development – not to mention natural causes such as floods and landslides – there were disappointments. Some of the trees Wilson photographed had disappeared, or if still present had been obscured by buildings. It was the same with structures; bridges lost or rebuilt, villages expanded or relocated. Fortunately, the authors found plenty of evidence remaining to record and photograph, enough to fill many gaps in our previous knowledge of Wilson's routes. One result is a splendid series of then-and-now images; these alone make their research and journeys worthwhile.

In addition, we have stunning portraits of 'Wilson' plants rediscovered by the authors along the way. I cannot look at those of the lampshade poppy *Meconopsis integrifolia* or the red flag poppy *M. punicea* without my pulse rate quickening; whilst those of *Rhododendron przewalskii, Rheum alexandrae* and *Primula secundiflora* among others tempt me to grab my boots, pack my bag and get me to those high places where Wilson first set eyes on them.

Flanagan and Kirkham succeeded in their quest for the spirit of Wilson in Sichuan and, as their book amply demonstrates, did so with the dogged determination that characterises all successful explorers. But it wasn't all grunt and grind and the sweat of endeavour. Their curiosity and sense of wonder accompanied them throughout, while their sense of humour made light of difficulties and disappointments. Like Wilson before them, they found that downs are usually followed by highs and that the end result is the satisfaction of a job well done.

An important beneficiary of the authors' work will be the plants themselves. Work has already begun on identifying and sourcing the least-known of Wilson's original tree introductions still in British cultivation, with a view to propagating and conserving their stock for future generations to enjoy. The great man would have approved.

Roy Lancaster
February 2009

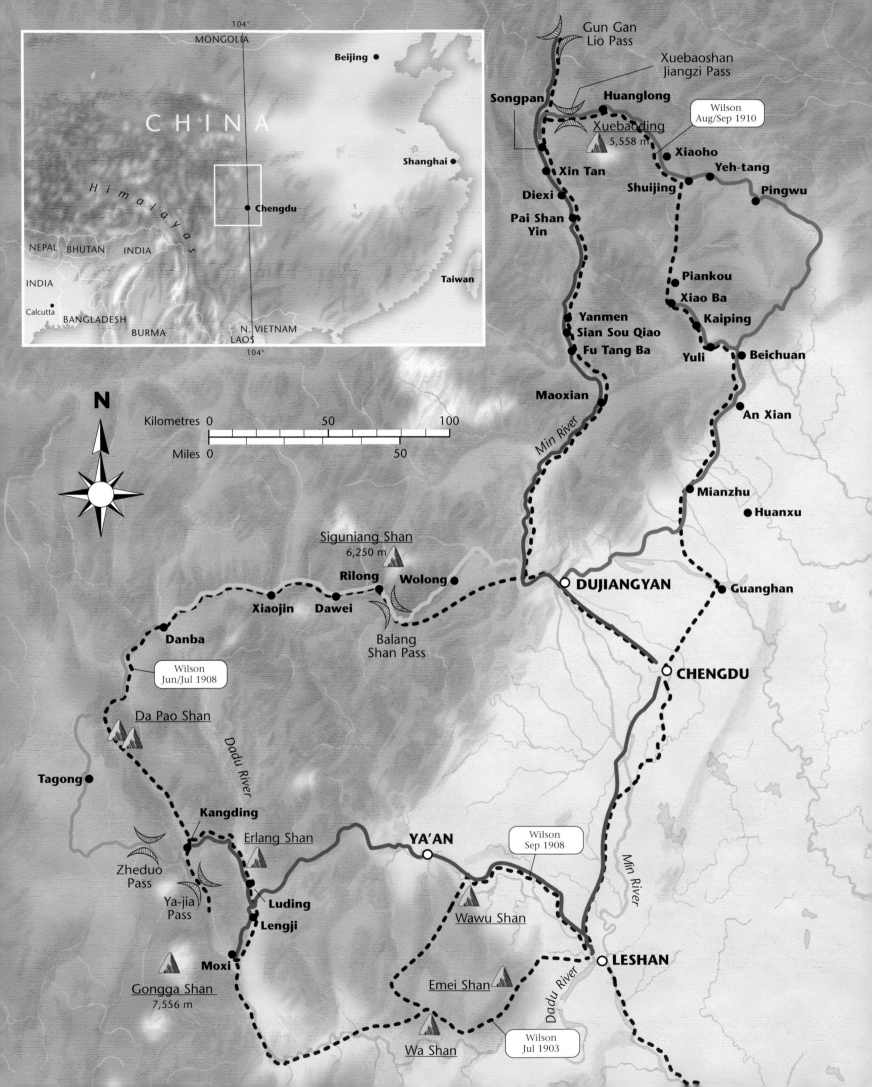

104°

MONGOLIA

Beijing

CHINA

Himalayas

Shanghai

Chengdu

NEPAL
BHUTAN
INDIA

INDIA

Calcutta

BANGLADESH

BURMA

N. VIETNAM

LAOS

104°

Taiwan

N

Kilometres 0   50   100

Miles 0   50

Gun Gan
Lio Pass

Xuebaoshan
Jiangzi Pass

Songpan   Huanglong

Wilson
Aug/Sep 1910

Xuebaoding
5,558 m

Xiaoho

Yeh-tang

Xin Tan   Shuijing   Pingwu

Diexi

Pai Shan
Yin

Piankou

Xiao Ba

Kaiping

Yanmen
Sian Sou Qiao
Fu Tang Ba   Yuli   Beichuan

Maoxian   An Xian

Min River

Mianzhu   Huanxu

Siguniang Shan
6,250 m

Rilong   Wolong

DUJIANGYAN   Guanghan

Xiaojin   Dawei

Danba

Balang
Shan Pass

CHENGDU

Wilson
Jun/Jul 1908

Da Pao Shan

Dadu River

Tagong

Kangding

Erlang Shan

YA'AN

Wilson
Sep 1908

Min River

Zheduo
Pass

Ya-jia
Pass

Luding

Lengji

Wawu Shan

Moxi

Gongga Shan
7,556 m

Emei Shan

LESHAN

Dadu River

Wa Shan

Wilson
Jul 1903

INTRODUCTION

In February 1911 the renowned plant hunter Ernest Henry Wilson left China for the final time. As he boarded the steamer bound for San Francisco he no doubt reflected on the previous 11 years since his first arrival in China. His legacy to history, as the foremost botanical collector of his generation, was firmly cemented. He had almost single-handedly transformed the gardens of the West with the exotic plant treasures of the East. And yet his situation was uncertain.

Wilson was leaving China as an invalid, his right leg had been broken in two places during an accident the previous September, and though he had received medical attention it was still heavily bandaged and causing him great pain. He also had personal and professional issues to deal with: his employment prospects were uncertain, his wife and young daughter had been domiciled, not always harmoniously, with his own family in England for just over a year and he was unclear as to what the future would hold.

Characteristically, at this time of uncertainty and following his brush with death, Wilson remained philosophical:

> The accident has put an end to my travels and rendered me a bit of a cripple to the end of my days, but I have been able to wind up my work successfully and in consequence I have no vain regrets. I have enjoyed my work in China and am proud in the knowledge that I have been privileged to achieve success in worthy employ and am certainly not going to pull a long face because the fates have unkindly given me a parting kick.[1]

Wilson was leaving a country which itself faced an uncertain future. Though China had endured centuries of turbulence and bloodshed, its central institutions, as represented by its imperial heart, had provided certainty and continuity. Now the power of the last emperor was broken and China's foes were gathering. Internal strife would soon consume the remnants of the Qing dynasty and end over two thousand years of dynastic rule.

Wilson would live for another 19 years and build on his international fame in the worlds of botany and horticulture, becoming an iconic figure to succeeding generations; whilst China would undergo a metamorphosis of philosophy and outlook, becoming the world's most populous communist state. In the middle of the twentieth century her doors firmly closed to the West as first the Great Leap Forward and then the Cultural Revolution convulsed the country and stretched her people to breaking point. The flower-rich provinces of western China where E. H. Wilson had trod the lonely paths and brought back an unimaginable haul of ornamental plants receded into history.

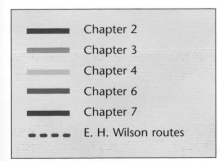

Chapter 2
Chapter 3
Chapter 4
Chapter 6
Chapter 7
E. H. Wilson routes

When China re-emerged onto the international scene in the late 1970s and early 1980s, blinking in the bright spotlight of western scrutiny, she would be a very different country. A systemic and state-sponsored attempt had been made to expunge her imperial past. All western influence had disappeared; the ruling communist party held the country in an iron-grip. The first tentative scientific exchanges began to reopen the mountainous areas in Hubei and Sichuan provinces where Wilson had plied his trade. But everything had changed. Though Wilson would have recognised the landscape it is sure that much else would have been unfamiliar to him. Many of the place names he had known had changed completely or were now transliterated in a quite different way. The rigid hierarchy that had defined Chinese society in Wilson's time had also been swept away, to be replaced by new structures and arrangements led by a new and ruthless ruling elite. Had he been reincarnated as a late twentieth-century plant hunter, Ernest Henry Wilson would have been a stranger in a strange land.

The landscape of rural Sichuan, a scene unchanged since Wilson's time despite the breakneck speed of modern China's development.

A charming portrait of the Wilson family. Mrs Wilson's demeanour seems to betray her general disregard for the trappings of fame her husband's activities evinced.

A plant label on an original Wilson collection growing at Kew. The Kew database uses a four-digit abbreviated code – WILS – to identify the original collector.

Mark and Tony in the forests of central Taiwan in 1992, an area that Wilson had previously traversed in the early twentieth century.

As western collectors re-entered the forests, upland pastures and alpine regions of China, attempts were made to ascertain exactly where Wilson and the other collectors of the so-called 'Golden Age' of Victorian and Edwardian plant exploration had travelled. In Wilson's case, apart from the best known locations, this proved to be very difficult and a consensus view developed that suggested reconstructing his journeys presented great problems.[2] E. H. Wilson had become an elusive hero.

My own awareness of the great man came with an increasing knowledge of garden plants during my horticultural apprenticeship. *The Hillier Manual of Trees and Shrubs* and W. J. Bean's great tomes, *Trees and Shrubs Hardy in the British Isles* – both key textbooks in my studies – were chock-full of references to Wilson's plant introductions. The wonderful gardens at the Royal Botanic Garden Edinburgh, where I undertook a diploma course in the early 1980s, contained a great many trees and shrubs embellished with the collection code 'Wilson' on their identifying labels, indicating plants raised from seed he had gathered, as did other botanic gardens and arboreta I came to know. In 1989 I was fortunate to visit South Korea on behalf of the Royal Botanic Gardens, Kew and in 1992 I journeyed to Taiwan.[3] On both these trips I was aware that I had, along the way, crossed Wilson's path and with a growing awareness of his exploits I became, like so many others, firmly captivated by the man. When the opportunity to visit China presented itself in 1996 I took it with both hands. The visit to Sichuan Province was all that I had expected it to be; the culture, people, landscape and flora were everything I had hoped for. But in this heartland of his work there was little hint or sense of E. H. Wilson. I became convinced by the received wisdom that he had passed into history.

Wilson's image of the famous Taiwan red cypress (*Chamaecyparis formosensis*) at Alishan, taken in 1917.

The same tree in 1992, now much reduced. Recognising this tree provided the first tantalising link to Wilson's time.

Kangding, in Wilson's day the important border town of Tachienlu, still nestling in its deep valley.

My third visit to Sichuan in 2001 was to be very different. In the company of Tony Kirkham, Head of the Arboretum at Kew, his colleague Steve Ruddy, an IT specialist, and William A. McNamara, the Director of Quarryhill Botanical Garden in Sonoma, California, we travelled in classic Wilson territory: west from Chengdu, the provincial capital, to Kangding, described by Wilson as 'a thriving trade entrepôt' and in his day a town of great political importance.[4] Then known as Tachienlu, Kangding was familiar to all the important collectors and travellers in the late nineteenth century and Wilson visited it many times, taking numerous atmospheric photographs of the town nestling in a deep valley between soaring, rugged peaks. Still I had little expectation of encountering any links with my hero, beyond experiencing the town that Wilson had known so well. All this was to change, however, when we moved to the village of Moxi, south-east of Kangding. Moxi also occupies a most striking location, again surrounded by high mountains, this time clad with thick forest and topped by permanent snow; the town essentially sits on a large island in the fast-moving Moxi River. Something so extraordinary was to happen that immediately on my return to England I wrote up the experience and sent it for publication in *Arnoldia*, a magazine published by the Arnold Arboretum, which was Wilson's employer following his return from China. I can do no better here than to reprint, verbatim, the key section of that article relevant to the present work. Having heard that a famous, though now dead, tree of the Chinese fir (*Cunninghamia lanceolata*), reputed to have been nearly 2,000 years old, was growing in the town we decided it shouldn't be missed:

after dinner we walked through the rapidly gathering twilight up the main street of old Moxi. As the tree's silhouette came into view above the surrounding buildings, I felt a strange sense of recognition. Then, when we cleared the last of the buildings, I blurted out, 'I know this tree'. As my companions knew I had never been anywhere near Moxi before they were rightly incredulous. 'I know this tree,' I repeated. 'It was photographed by Wilson.' The characteristic shape and distinct leaning posture of the tree made me certain that it was the same one I had seen in a photograph by Wilson in W. J. Bean's Trees and Shrubs Hardy in the British Isles.[5]

It didn't prove difficult to confirm that this was indeed the same tree that Wilson had photographed as a flourishing specimen on 17 July 1908. Though many western people had visited Moxi since the mid-1980s I couldn't find any evidence that anyone had connected this dead Chinese fir to Wilson's photograph. For me this provided the first direct link with Wilson's China and began a quest that was to lead to this book and the journeys it describes.

Fortunately, I had a confederate in my quest, a person with an equal fascination with Wilson and a determination to try and recapture his world: Tony Kirkham. Tony had been responsible for starting me on my own plant-collecting career, when he selected me to accompany him to South Korea in 1989, where he led an expedition to collect seeds for replanting Kew's Arboretum following the devastating effects of the Great Storm of 1987. This trip led to visits to several other countries across eastern Asia, as discussed in our joint book *Plants from the Edge of the World*.[6] During this time we shared the highs and lows that accompany plant-collecting activities and began to develop a deep appreciation of the work of these Golden Age plant hunters. But most of all we began to develop a particular affinity with E. H. Wilson, who became our exemplar and invisible companion. At times it seemed that whatever we did on our travels – visiting the remote island of Ullung Do in

The front cover of *Arnoldia*, the magazine of the Arnold Arboretum. Mark's article detailed the 'rediscovery' of Wilson's Chinese fir. This important milestone set the authors on their quest to trace Wilson's journeys.

'*Cunninghamia lanceolata* south west of Tachien-lu.' The largest specimen that Wilson ever saw.

Wilson's tree was killed by a catastrophic fire in the 1990s and is today a shell of its former self. Despite these vicissitudes its attendant temple, also shown in Wilson's image, still remains in a modified format.

Mark and Tony with Peter Gorovoy celebrating the conclusion of a successful plant collecting expedition on Mount Chekhov, Sakhalin Island in October 1994.

the Sea of Japan, climbing to the top of Yu Shan, Taiwan's highest peak, walking amongst the rare Korean firs – Wilson had been there and done that. It was difficult not to be in awe of such an individual.

The visit to Moxi in 2001 was followed by another trip to China in 2003, centred around Songpan. A key aspect of this trip was to confirm the realisation which had been growing in our minds that, backed up by research and a degree of luck, it might actually be possible to accurately stitch together Wilson's Chinese journeys. This seemed all the more appropriate given that the first decade of the twenty-first century marks the centenary of Wilson's time in China. However, our hope wasn't just to retrace Wilson's journeys, the trip also offered other fascinating opportunities.

The passage of time has wrought profound changes not just in China, as previously discussed, but in Wilson's legacy to western gardens. Many of the plants he introduced became firm garden favourites propagated innumerable times and still freely and cheaply available at any garden centre. Others, however, did not. Consider the case of *Meliosma beaniana*, a deciduous tree much admired by Wilson: 'The pinnate-leaved members of this small family are all handsome trees and none was in cultivation previous to my explorations. I have succeeded in introducing three species, all of them promising to thrive under cultivation. One, *M. veitchiorum*, is now flourishing just within the main entrance to Kew Gardens.'[7] *Meliosma beaniana* was named for his great friend William Jackson Bean, the Curator of Kew. Wilson introduced it from western Hubei in 1907. In all his travels he only encountered it on a handful of occasions; there is no place where this is a common tree. Today only two specimens exist in cultivation, one at the Royal Botanic Garden Edinburgh and one at Caerhays Castle in Cornwall. And this is by no means an isolated example. The collections of the RBG Edinburgh, along with those of the Arnold Arboretum, are recognised as being the repositories of the most extensive collections of authentic, first generation Wilson plants in cultivation; and yet the Edinburgh Living Collection contains only 161 extant taxa,[8] a small representation of the estimated 1,200 plants, and these trees and shrubs only, that he introduced.[9] Clearly Wilson's legacy is dwindling. We thought the visit to China might, therefore, also offer the opportunity to assess the status of some of Wilson's key collections. How had they fared in the intervening ten decades?

The associated issue of the condition of the forests that Wilson had journeyed and collected within, and the degree to which deforestation has affected them, was also of considerable interest. Environmental degradation in China is an issue of the utmost complexity. For millennia the growth in Chinese civilisation and population took a heavy toll on the natural resources of the country. The lower-lying eastern and southern provinces lost their forests in antiquity but in the mountainous west much remained until the modern era. However, all the early collectors and travellers in the Chinese/Tibetan uplands make reference to the heavy hand of humanity. As early as 1876 Armand David, the pioneering French missionary, lamented that: 'From one year's end to another, one hears the hatchet or the axe cutting the most beautiful trees. The destruction of these primitive forests, of which there are only fragments in all of China, progresses with unfortunate speed. They will never be replaced.' Wilson himself was only too aware of the historical antecedents of the problem. Commenting on the diversity of the Chinese flora he tells us:

> *This extraordinary wealth of species exists, notwithstanding the fact that every available bit of land is under cultivation. Below 2,000 feet the flora is everywhere relegated to the roadsides, the cliffs and other more or less inaccessible places. It is impossible to conceive the original wealth of this country, for obviously many types must have perished as agriculture claimed the land, not to mention the destruction of forests for economic purposes.*[10]

The references in recent literature to a continuation of the felling are legion, with some in exactly the areas where Wilson travelled. Those of the great American biologist, George Schaller, are perhaps the most pertinent and depressing.[11] But what might a direct comparison of Wilson's landscapes with the present day situation reveal? Though we had no intention of joining the debate formally we felt it would be interesting to make some pertinent observations, which would add a further strand to our project.

In addition to visiting China a parallel exercise also seemed to suggest itself, given the parlous state of many of Wilson's plants in gardens. Had anybody undertaken a systematic stocktake of his remaining collections in gardens and initiated a propagation programme? Apparently not. In the summer of 2005 Tony and I coordinated a workshop at Kew, under the auspices of the PlantNetwork of Britain and Ireland, to which all the key gardens and arboreta were invited. During a series of lectures we explored the nature of Wilson's legacy in both China and UK gardens. At the conclusion of this workshop the collective feeling was that a propagation initiative was worthwhile and that a journey to revisit several of Wilson's key sites in China would also be of great interest. Buoyed up by this endorsement Tony and I pressed on.

Tony speaking at the Wilson workshop at Kew. Behind him is the original *Catalpa fargesii* tree WILS 664, collected by Wilson in Hubei.
Photograph:© Mark Flanagan

Ian Bond, our principal sponsor, at home on the fells of his Highland home – Morar – north of Fort William.

Organising such a trip, so easy to propose, would not be straightforward; it would involve time, money and organisation. Fortunately we had the necessary contacts in China and a range of interested sponsors in England. The latter were led by Ian Bond, a keen plantsman and gardener with one of the most comprehensive private collections of walnuts – species and cultivars of the genus *Juglans* – which is now registered as a National Collection® by the National Council for the Conservation of Plants and Gardens. Indeed, Ian's interest extends to other members of the family Juglandaceae and woody plants in general. Ian's home and beautifully landscaped garden is situated in the heart of the Gloucestershire countryside, close to Chipping Campden, the birth place of Wilson. With his foresight and close affinity to Wilson, Ian saw the historical value and close links of our proposals. He not only put his own money into the project but persuaded three of his close friends, Michael Heathcoat Amory, Lord Dulverton and Lady Arabella Lennox-Boyd to do the same. With this financial backing we were able to move to the next phase. Our collaborators in China were old friends from our expeditionary work who were also enthused by our interest in Wilson. Dr Yin Kaipu, in particular, had developed a fascination with Wilson and is, probably, the foremost Chinese expert. However, certain aspects of Wilson's journeys remained unclear and there was only one way to clarify matters: we had to visit the Arnold Arboretum in Boston in the United States.

The majority of Wilson's papers are deposited at the Arnold Arboretum, his employer at the time of his death in 1930. This archive of information is expertly curated by the skilled and dedicated staff in the Arboretum, part of the Library Services of Harvard University. Once again, Tony and I were fortunate to have on-site assistance. Dr Peter Del Tredici was, at the time, Director of the Living Collections at the Arnold Arboretum and is the modern successor to Wilson himself. Not only is Peter an accomplished botanist, he is also a man of wide interests with a challenging and radical intellect, and an amusing raconteur to boot. This combination led to Peter undertaking a tongue in cheek experiment which discovered that the Watergate hearings, which brought about the impeachment of Richard Nixon, depressed the growth of plants! He established this by exposing pea seedlings to a radio broadcast of the hearings, and those subjected to the mind-numbing detail of this complex case grew less well than those that were not![12] More seriously, Peter is a fellow student of Wilson and provided us with direct help when we visited the Arnold Arboretum in October 2005.

Mark in the library of the Arnold Arboretum, transcribing information from Wilson's field journals. By cross-referencing these notes with herbarium material at Kew and Wilson's photographic record, it became possible to accurately determine Wilson's journeys.

Over the course of several days in the arboretum's library we were able to expand our knowledge of Wilson's journeys, by direct reference to his field journals and correspondence, to the extent that it became possible to set out an itinerary for a visit to China that would follow, in detail, his travels in the summers of 1908 and 1910 and part of his first foray into Sichuan in 1903/04. Armed with this additional information we could begin our quest to find Mr Wilson. Our journey would not break new ground, a great many collectors and travellers had visited the areas we intended to visit, some with an awareness of Wilson's activities, though most without. What would make this trip unique and valuable is that Tony and I intended to make tracing Wilson's travels the central purpose of our visit to China. This was not a trip to collect plants, as all our previous visits to China and the Far East had been; we would not be encumbered by the need to collate field notes or gather herbarium specimens. This visit would focus solely on finding the places E. H. Wilson had visited by reference to his writings and by a direct comparison of the many photographic images that he took during his journeys. In this way we hoped to gain a real sense of the man and his times. It promised to be a journey of more than ordinary interest.

CHAPTER ONE

ERNEST HENRY

A great deal has been written about E. H. Wilson. The key facts of his life – his birth in the lovely Cotswold town of Chipping Campden on 15 February 1876, his early professional life in the famous nursery of Hewitt's of Solihull, his entry into the world of botanic gardens, firstly at the Birmingham Botanical Garden and later at Kew, his great achievements in China and other eastern countries, his executive role as Keeper of the Arnold Arboretum in Boston and finally his untimely death at the age of 54 on 5 October 1930 – are widely known. And yet he remains elusive. The key biographies that chart his life reveal surprisingly little about the man himself and are largely disappointing in helping us to really get under the skin of Wilson in a way that other biographies of Edwardian adventurers have done.[1] It is a great pity that Wilson has not been subjected to the same learned treatment that has been given to Robert Scott, for example, where the core of the man, his flaws, weaknesses, strengths and virtues are exposed by the objective rigour of the scholar.[2] We can find no such analysis of E. H. Wilson by either of his biographers. Not that their books are hagiographies, indeed both recognise that Wilson was no saint, readily acknowledging his many shortcomings; his occasional irascibility, jingoism and narrow, almost prudish, essentially Victorian outlook. We also know very little about Wilson's personal relationships, particularly with his wife Ellen (Nellie), who remains a rather shadowy figure, though anecdotal and documentary evidence tells us that she was never reconciled to Wilson's long and frequent absences nor, indeed, to life in the United States.

Interestingly, some of Wilson's contemporary collectors have received recent attention with Brenda McLean's biography of George Forrest giving us greater insights into his life,[3] and the reworking of Frank Kingdon-Ward's *Riddle of the Tsangpo Gorge* by Kenneth Cox, which provides new and revealing information about his important 1924–26 expedition to south-east Tibet.[4] A little gem of a book by Nicola Shulman has also given us a very different perspective of the obdurate Reginald Farrer.[5]

In writing an account of Wilson's remarkable journeys in China we need to understand the man, because by understanding the man we can understand what motivated him to take such a life path. Where do we turn to discover the real Ernest Henry? Roy Briggs had access to more of Wilson's papers than anyone, including family correspondence that is more intimate and unguarded than his inevitably businesslike and anodyne official papers, and yet even he could only find hints of the real Wilson. Edward Farrington knew Wilson in life and could call himself a friend and yet it seems that he too failed, or

A studio photograph of Wilson in England, probably taken as his fame was beginning to spread following his return from the second Chinese expedition.

E. H. Wilson was born in this Cotswold house to Henry and Annie Wilson.

preferred not, to penetrate below the surface; he provides us with a series of clichés rather than objective observations.[6] It seems to me that consideration of Wilson 'the man', has been overwhelmed by Wilson 'the legend'. The scale of his achievements and his towering reputation have prevented a more rounded consideration of Wilson as a human being with all the usual insecurities and weaknesses that we are all subject to. I do not intend to undertake such an approach here but I have not been able to pursue my own story – finding Wilson in China – without gaining some personal insights into Wilson's character and outlook. Much of what I may say on this subject is, I freely admit, conjectural, being the result of an intangible 'feel' for Wilson gained almost empathetically by being where Wilson stood.

The cold, windblown Ya-jia Pass is one such place, high in the Da Xue mountains to the west of Kangding, where historically China met Tibet. At times it can be as god-forsaken as anywhere on the planet. Wilson visited this place on at least two occasions. Amongst the shattered rocks and in the rarefied air, a wonderful thing happens each spring. The seasons are short here, nearly 4,000 m above sea level, but as the snow melts at the end of May beautiful flowers emerge from the turf, some of the most intriguing blooms that nature has to offer. One such is *Primula amethystina* (p. 30), just a few centimetres in height but with dainty flowers of an elusive colour, not truly blue but without any hint of pink. Where the lie of the land offers a modicum of protection from the incessant wind the lampshade poppy (*Meconopsis integrifolia*) unfurls its rich, primrose-yellow flowers. It was from below this pass that Wilson collected this species, the main object of his second visit to China, on 16 July 1903. Other poppies – red-flowered *Meconopsis punicea* and azure *M. henrici* (p. 31) – join the display. By August it is all over and in September the first flakes of snow begin to fall before the whole scene is blanketed for six months in a white mantle. Wilson took a memorable photograph here on 19 July 1908, and on 17 June 2006 Tony Kirkham and I stood in exactly the same spot marvelling at a view which had changed not one jot in the intervening 98 years.

In such a situation Wilson's feelings become almost palpable and a sense of kinship develops despite the passage of time. It is easy to gauge Wilson's state of mind – his discomfort and fatigue with perhaps a hint of apprehension – but, overwhelmingly, his emotions would be wonderment and excitement. The scenery is sublime and few, if any, non-Chinese or Tibetans had passed this way before him. It was *terra incognita*. The boy from the Cotswolds, who proved himself a more than able student, caught the eye of the Director of Kew and was dispatched to China in search of the fabled dove tree, feels a sense of personal pride that few would begrudge: here in the mountains of China his destiny was being fulfilled.

At the World Heritage Site of Dujiangyan – Kuan Hsien to Wilson – we find a different, more reflective man. Here on the edge of the Chengdu Plain is a famous irrigation works, of such importance that it sits close to the apex in the pantheon of Chinese achievements. Li Bing, the Governor of Sichuan Province during the turbulent Warring States period of Chinese history, devised a means to harness the Min River and provide ample and dependable water supplies to the plain below with the result that it became one of the most fertile areas in China, a veritable rice bowl. Wilson passed through Kuan Hsien many times and was clearly impressed and indeed moved by what he saw. He writes:

> *The originators of this work have been deified and two magnificent temples overlooking their work at Kuan Hsien bear witness to the gratitude of the millions who have enjoyed, and continue to enjoy, prosperity from the labours of the famous Li Bing and his son. The 'hero-worship' here exemplified would do credit to the people of any land.[7]*

That Wilson had a genuine regard for the Chinese people is indisputable. It was a point that was noted during his lifetime and, remarkably, it is reflected to this day in the attitude of my Chinese friends.[8] The Han Chinese are a complicated race. One of the four great civilisations that ushered in the modern era, along with the Egyptians, Mesopotamians and the Indus civilisations, for centuries they regarded themselves above all other races.

The Da Xue Shan is a mountain range with tremendous snow-clad peaks and spurs which act as the eastern bastion to the Tibetan Plateau.

The delicate flowers of *Primula amethystina* belie the rigorous environment in which it grows.

They are the most populous race and amongst the most vibrant on earth, and yet they have deep insecurities. The certainty of their values, forged over millennia, crumbled in the face of nineteenth-century western imperialism which left a deep legacy of mistrust and injustice. The collective memory of imperialism is not a happy one and many of the key western figures are regarded with ill-concealed contempt. Yet Wilson is not only remembered as a figure of great interest but one who evinces a certain degree of warmth. Those Chinese who know of Wilson, though admittedly they are few, regard him with a partiality offered to few of his contemporary collectors, a consequence of their perception that Wilson treated his Chinese team with kindness and respect at a time when most westerners were openly racist in their behaviour. Wilson also showed respect and reverence for the social graces and institutions of Imperial China: in the words of Frank Meyer, who collected in China at the same time as Wilson, he 'likes to pay his respects to the hundred and one officials.'[9] This warm regard on the part of modern Chinese people strikes me as most significant and indicates a positive aspect of Wilson's character that does him much credit. In his writings it is easy to find a great many favourable comments about the people he worked and associated with:

> In the spring of 1900 I engaged about a dozen peasants from near Ichang. These men remained with me and rendered faithful service during the whole of my peregrinations. After a few months training they understood my habits thoroughly and never involved me in any trouble or difficulty. Once they grasped what was wanted they could be relied upon to do their part, thereby adding much to the pleasure and profit of my many journeys. When we parted in February 1911 it was with genuine regret on both sides. Faithful, intelligent, reliable, cheerful under adverse circumstances and always willing to give their best, no men could have rendered better service.[10]

A more ringing endorsement is difficult to imagine.

He is equally magnanimous about the many minority people he encountered: 'Though my associations with the Sifan were brief I always received the utmost courtesy at their hands and found much that was pleasing and interesting among these happy, unsophisticated children of Nature.'[11] Such an attitude is remarkable at a time when the concept of the superior white race was not only universal in Europe and North America but virtually the state policy of His Majesty's Government.[12] This is not to say that Wilson was above disparaging people but his derogatory comments are few and largely aimed at specific individuals, the result of frustration or some perceived slight. Wilson's writing is almost completely devoid of the type of distasteful remarks made by Frank Kingdon-Ward, for example, whose stereotyping of native people is notorious.[13]

What then do we make of Ernest Henry Wilson, having followed in his footsteps? What kind of man emerges across the years that separate his time from ours? Can the journeys we recreated offer insights into the character of Wilson denied to others? First and foremost a very human person emerges, a person stripped of the aura that history wrapped around him, a man of great achievement but a man nevertheless. He was subject to ill health:

> *During the night at San-tsze-yeh I had a violent attack of ague, probably caused by a chill which culminated in a fit of vomiting. This seizure and the howling of many dogs were against a good nights sleep. In consequence we took things very gently the next day and I used my chair much more than usual.*[14]

Frustration, fatigue and a loss of heart are the stock in trade of the plant hunter and Wilson was no different, 'I am certainly getting tired of the wandering life and long for the end to come. I seem never to have done anything else than wander, wander through China.'[15] But we also find much to inspire us, which accounts for Wilson's enduring reputation nearly a century after his epic Chinese journeys. Having travelled the roads that Wilson took we can attest to his endurance and fortitude, having surveyed the plants that Wilson saw we can confirm his skill and expertise; but most of all we can reaffirm that history has accurately reflected the breadth and depth of his achievement and rightly placed him highest in the constellation of the great plant collectors.

This is the Ernest Henry that we found, amongst the teeming cities, expansive plains and towering mountains of western China, during our journey to find Mr Wilson and reconnect with the China that he knew.

Meconopsis henrici var. *henrici*, a solitary-flowered species found throughout the mountains of western Sichuan. This charming blue poppy was discovered by Prince Henri d'Orleans after whom it is named.

CHAPTER TWO

WAWU – LUDING, DADU – KANGDING

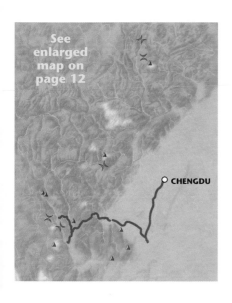

See enlarged map on page 12

CHENGDU

A bronze statue of a tea porter by the roadside outside Ya'an (see p. 45).

Our adventure began in the provincial capital of Chengdu. The Air China flight from Beijing had been uneventful and this stage of the journey was very familiar to us. We had been to Chengdu many times as part of a Kew-led consortium of plant collectors and the welcoming party that greeted us in the airport lounge was just as familiar. Yin Kaipu, a retired ecologist, and Zhong Shengxian, the Director of the Library, are both associates of the Institute of Biology, part of the China's principal national scientific organisation Academia Sinica. Both these men are good friends having personally supported all of the Kew trips stretching back to 1988. The smiles, hugs and backslapping were a tangible demonstration of the depth of our friendship. The two drivers present were also old friends. The more senior driver, Wang Hangming, was to prove invaluable on this very different trip. This time there would be no seed collecting, no gathering of herbarium specimens and no recording of field notes: our primary interest wasn't plants at all but the routes followed by Wilson in the early decades of the twentieth century. We had decided that as no seed collecting would take place we would travel in midsummer. In all our previous visits to the East we had never had the opportunity of seeing plants in flower. In seeking out seeds and fruits it is necessary to travel in September and October, long after many plants have flowered. On this occasion we would be able to see some of nature's most spectacular floral displays during their prime season.

Wilson also knew Chengdu very well, he visited it on many occasions at different times of the year and writes with obvious interest and affection for this ancient place. He was also aware of its great antiquity. Chengdu was founded over 2,000 years ago and has been an important regional centre since that time. Its importance is accounted for by the position it occupies within the province, more or less equidistant to the other major centres of population such as Chongqing and Langzhou. It also bestrides the road that leads north into Shaanxi – a most important highway since pre-dynastic times – and it lies at the head of the plain that provides the agricultural wealth of the province.

Wilson's journeys to Chengdu were very different to ours. The three hour flight from Beijing is a product of the late twentieth century. Travellers in past years had to climb the high passes from the west using the road from distant Lhasa, or journey down the Yangtze River to Yibin and on to Leshan along the Min River before continuing overland. Rivers were a fast and effective, though dangerous, way to gain access to the interior of the vast Chinese Empire at a time when roads were rudimentary or non-existent. Wilson hired his own river boats or used the services of local boatmen, a group of men for whom he had the greatest respect: 'These Chinese boatmen are careful, absolutely competent and thorough masters of their craft, and the more one sees of them and their work the more one's respect grows. Oriental methods are not Occidental methods, but they succeed just the same!'[1]

We didn't dwell in Chengdu, a quick spot of lunch and we were on our way. The Institute had two Mitsubishi four-wheel drive vehicles which could be hired for journeys into the more remote parts of Sichuan. Soon we were speeding down the expressway to Leshan, a distance of 162 km (100 miles). Wilson knew Leshan as Kia-ting Fu and used it as a regular base, particularly in winter when he relaxed after the exertions of the rest of the year and organised his affairs between collecting seasons. We arrived within two hours and went straight to our first site. Wilson took an interesting image of the city walls at Leshan, a view which shows temple buildings crowning a hill. Dr Yin had enlisted some

Tony and Mark with porters and guides at Chaoke during a Kew expedition to the Gongga Shan area in 2003.

Our long-standing guides and friends, Yin Kaipu and Zhong Shengxian from The Academy of Sciences in Chengdu.

Despite our misgivings, a closer inspection of the buildings confirmed that we were on the site of Wilson's temple image.

local support; Mr Hu was a resident who knew the city well and would lead us to the spot where Wilson had taken his image. On arrival it was clear that things had changed dramatically (see following pages) and Tony and I were less than convinced we were in the right place.

Trying to match the view meant a hike across the hillsides, through the exercise ground of a local technical college, up the stairs of a dormitory building and onto the roof. This rigmarole was to become familiar in built-up areas, a consequence of the development which has occurred in the intervening decades. Even after all this effort we still weren't convinced we were at the same place. A climb up the hill Wilson photographed and into the grounds of the temple provided more evidence that our guide had correctly identified the area. Close up the details of the temple building could clearly be matched to Wilson's image, but it was hardly the auspicious start that Tony and I had hoped for and we were left rather deflated – this was going to be more difficult than we had envisaged.

32. Kia-ting Fu, W. Szech'uan. View of temple on city wall. 1,100 feet.
September 1, 1908.

'Kia-ting Fu, W. Szech'uan. View
of temple on city wall. 1,100
feet. September 1, 1908.'

The same view in 2006. Not surprisingly it became apparent that locations in urban areas had undergone much greater changes those in rural or mountain areas.

34. Kia-ting Fu, W. Szech'uan. Sandstone cliffs facing junction of the Tung and Min Rivers, looking north. 1,100 feet. Nov. 28, 1908.

'Kia-ting Fu, W. Szech'uan.
Sandstone cliffs facing junction of
the Tung and Min Rivers, looking
north. 1,100 feet. Nov. 28, 1908.'

Ninety-eight years later the scene is similar, although there appears to have been a marked deterioration in the quality of the air and the river levels are much higher.

The famous Leshan Buddha.

Leshan is famous as being the city at the confluence of three rivers, two of which are amongst Sichuan's great rivers, the Min and the Dadu, the latter known to Wilson as the Tung River. At just this point a huge Buddha was carved into the rock face during the reign of the Tang Emperor Kaiyuan. Work began in 713 AD and took more than 90 years to complete; not surprisingly this is now one of Sichuan's many World Heritage Sites. Surprisingly, Wilson does not seem to have taken a photograph of this amazing carving, nor does he make reference to it in his writing. Wilson's second image in Leshan, the one we hoped to match, was taken literally just around the bend in the river from the giant Buddha and shows a view across the confluence point to a forested hill where an enigmatic pagoda sits amongst the trees. This image was altogether easier to match and is little altered, though the pagoda has been somewhat modified. The key difference we were faced with was the height of the river. Wilson took his photograph at the end of November, when the water is at its lowest. The winter months over much of temperate Asia are dry with little rainfall, additionally the headwaters of the Dadu and Min rivers are located high in the mountains and are locked under a blanket of snow and ice. As a consequence, Wilson's image shows gravel bars and rocky shoals and temporary habitations close to the river. In contradiction, June is the wettest time of year with abundant rainfall and meltwater rushing down from the mountains. We were confronted with fast moving and obviously rising rivers. Their waters formed eddies and whirlpools at their point of confluence and the strength of the current was easy to gauge by the difficulties experienced by the many swimmers who braved the water. All these rivers can rise very fast and unpredictably and in the past have been responsible for great loss of life, even in quite recent times.

A little more encouraged and with the start of a sense of having stood where Wilson stood we spent a comfortable night in Leshan and over dinner reconfirmed our plans with Dr Yin. We would drive north-west towards Wawu Shan (Tiled House Mountain, an illusion to its supposed resemblance to the flat-topped, tile-roofed houses of ancient China), one of Sichuan's trio of isolated sacred peaks, following the journey made by Wilson in 1908. Then we would move on to the important city of Ya'an, now, as in Wilson's day, a centre for the tea industry. Our journey would continue west into the valley of the Dadu River to Luding, where an iron bridge passed into legend as the scene of a famous battle victory by Mao Zhedong's Red Army, and through the mountains to Kangding, Wilson's 'Gateway to Tibet'. This would mark the first stage of our journey and cover significant parts of Wilson's second trip to China between 1903 and 1904.

The following morning we were on the road early. The way from Leshan to Ya'an ran along the busy main road at first, though quite quickly we turned off and entered the broken country that Wilson knew as the 'laolin': 'The entire absence of decent roads, the sparse population, wretchedly poor accommodation, the savage cliffs and jungle clad mountainsides sufficiently entitle this region to be called "Laolin" i.e. "Wilderness."'[2] Though Wilson travelled in this area in 1903, including ascents of both Wa Shan and Emei Shan, the two neighbouring peaks, he didn't climb Wawu Shan until the summer of 1908 when he journeyed across the laolin from the north-east to the south-west, eliciting the above remarks. Things hadn't changed a great deal and though the area is more populated it certainly didn't appear to be any more prosperous. We had an image to match, an atmospheric view of Wawu Shan, a misty flat-topped mountain rising in the distance above a deep, river valley. Wilson's writing once again accurately describes the scene: 'From the summit of Tsao shan we obtained our first view of the Wa-wu shan, an extraordinary-looking massive mountain, singularly like Wa shan in contour, resembling a huge ark floating above the clouds of mist.'[3] Try as we might we could find nowhere that resembled this image, even with the help of local people, and it was difficult to establish whether we were actually on the same route as Wilson. A recent hydroelectric scheme in the river valley did little to help, the spoil and associated engineering works had caused a degree of general devastation. We continued on towards the mountain, the dilapidated temple at Tsung-tung-che where Wilson endured a flea-bitten night proving elusive. By mid-morning we had arrived at the cable car station that now conducts visitors up to the summit of the mountain. It was all but deserted and it is difficult to believe that this facility is a thriving concern, though it claims to serve the needs of tourists from Chengdu. Nonetheless, it afforded us a speedy and convenient way to the top, far removed from the slog that Wilson was forced to conduct: 'We dragged ourselves upward by grasping shrubs and it was a marvel to me how the coolies with their loads managed to overcome the ascent. The foothold was precarious and it was often a case of one foot forward and two backward!'[4]

Despite the help of friendly local people, we were unable to discover Wilson's route into Wawu Shan.

Ascending to the summit of Wawu Shan above the rich mixed vegetation. The silver firs on the ridgeline are *Abies fabri*.

Wilson regarded Wawu Shan as a relative disappointment. For collecting purposes he found little to whet his appetite, writing that: 'From a botanical point of view Mount Wawu proved disappointing'. The general condition of the mountain saddened him due to the large-scale felling which had fuelled the local iron smelting industry: 'all the mixed timber has been felled for making charcoal and other purposes, leaving only a dense shrubbery in which variety is not great'.[5] Though our visit was so fleeting as to prevent even a cursory estimate of species variety, it was certainly clear that a great deal of natural regeneration has occurred in the intervening years. This was our first brief chance to assess a most vexatious question: had deforestation continued unchecked since Wilson's time? Clearly the iron smelters had long since left Wawu Shan and this was most probably the key reason for the renewed vigour of native plants. However, Tony and I began to see another most important factor. Virtually everywhere we travelled in western and northern Sichuan we came across evidence of the abandonment of large-scale forestry operations. This was first-hand evidence of the success of a key state-sponsored conservation initiative: the Natural Forest Protection Program (NFPP). The project was a direct result of near catastrophic flooding of China's two main rivers, the Huang He (Yellow River) and Changjiang (Yangtze River), in 1998, and other environmental disasters such as the severe sand storms that engulf Beijing from the north-west from time to time. The undeniable conclusion of the inquiries which examined these crises and the opinion of external experts was that deforestation in the headwaters of the two rivers in the western mountains was causing uncontrollable flooding and desertification. The NFPP was the Chinese government's response to dealing with these issues. The programme established ambitious targets to reduce commercial logging in 17 provinces in the higher and middle

The summit of Wawu Shan is little changed from Wilson's time.

reaches of the Huang He and the Changjiang by 12.4 million m³ each year, with an associated afforestation project to plant 8.67 million hectares of new forest areas by 2010. A good deal of debate surrounds the success and effectiveness of these measures but we certainly saw several sites where tree felling had been curtailed. A second equally ambitious environmental initiative was launched in 1999: the Sloping Land Conversion Program (SLCP), with a budget equivalent to more than US$40 billion. The SLCP covers an even larger area with 25 provinces involved. Its central purpose is to withdraw marginal farmland, mainly above 2,000 m in elevation, from production and allow natural regeneration to take place to provide these areas with forest cover. This involves a massive restructuring of the rural economy affecting millions of people. Once again there is a degree of scepticism amongst the international community about whether this programme can meet its stated targets. Time will tell, but clearly these two associated projects offer the best chance that the unending exploitation and degradation of China's unique forests can finally be arrested for the first time in recorded history.

The vegetation below us, as we pitched and rolled in the cable car towards the summit, was composed of a mixed assemblage of trees and shrubs. Tall specimens of *Rhododendron calophytum* var. *openshawianum* could be seen, and in the understorey, the rugose leaf surface of another rhododendron easily identified as *R. wiltonii*. Rowan trees, birches and hornbeams were common, with an occasional evergreen member of the Fagaceae – perhaps a species of *Lithocarpus* – being evident. The journey was an exhilarating one and as we climbed higher we were able to look back to the Chengdu Plain lying below the attendant cloud cover. The mountain also has considerably more intrinsic interest than Wilson suggests, for example, it is noted for several picturesque waterfalls including the famous Lanxi, which falls over 1,000 m in height by a series of steps.

The summit was soon reached and we picked up a trail heading across the flat expanse of bamboo. 'The mountain top is undulating park-like and covered with an impenetrable jungle of bamboo-scrub,' was Wilson's accurate description.[6] He also noted the preponderance of Delavay's fir (*Abies delavayi*) – though modern opinion identifies the trees as *A. fabri*, which is also found on Wa Shan and Emei Shan. Being a sacred mountain Wawu Shan had various temples most of which were in a poor state of repair or, indeed, derelict during Wilson's visit, though he did lodge overnight at an extant temple building at Kwanyin-ping. We continued across the plateau-like summit for several hours, most of which were spent

'Tachien-lu, W. Szech'uan. Men laden with "Brick tea" for Thibet. One man's load weighs 317 lbs., the other's 298 lbs. Men carry this tea as far as Tachien-lu, accomplishing about 6 miles per day over vile roads. July 30, 1908.'

trudging through the unrelieved bamboo brake. A ruined temple was spotted in the distance but Xiao Zhong insisted it was certainly not the one where Wilson spent his night's sojourn. With little else of interest we returned to the cable car and back to the waiting vehicles, our visit to Wawu Shan proving to be of a rather academic interest. Within a few hours we were safely lodged in the Yu Du Hotel in Ya'an.

Ya'an, in Wilson's day Yachou Fu, is still an important provincial city and the clearing house for the export of brick tea to Tibet. Wilson tells us that, 'to the Tibetans tea is an absolute necessity of life, and deprived of this astringent they suffer in various ways'.[7] He devoted a complete chapter to the growing and export of tea in his *Naturalist in Western China* and he makes numerous other references to it in his various writings. Clearly this was something that fascinated him and with good reason. The Silk Road across the arid landscape of central Asia, is both well known and evocative. Historians acknowledge its enormous significance in spreading the products and knowledge of Chinese culture to the West. The so-called Tea-Horse Trail, by contrast, is much less well known and has received little attention from scholars. Nonetheless, it too had a significant role in linking western China with the tribes and petty kingdoms of Tibet, thereby spreading Chinese influence. What most caught Wilson's attention was the fact that the tea was carried to Kangding over the roughest terrain on the backs of porters (see previous page), a distance of 225 km (140 miles). The tea was pressed into brick-shaped 'chuan' and stacked to make each individual porter's load. Wilson himself had estimated loads of 168 kg (370 lbs)

A modern warehouse in Ya'an showing that the tea is still shaped into conveniently-sized brick 'chuan'.

being carried. Such superhuman efforts deeply moved Wilson, as did the fact that the porters were paid a pittance to undertake such work. This method of transport continued until the construction of the Sichuan–Tibet highway in the 1950s when the tea started to be transported in lorries, although it is still shaped into bricks as we saw when examining one of the many warehouses that line the roads into Ya'an. The last survivors of this trade are still alive to tell their tales. Many live in Ganxipo, a village along the trail a few kilometres west of Ya'an. One veteran, Li Zhongquan, remembered the ascent of Erlang Shan being the most demanding section of the journey: 'One misstep and you were gone. We'd had our sandals soled with iron to get over the mountain'. [8] The sheer weight of the loads compelled the porters to stop for a rest every few hundred meters. At the base of each load a wooden stick with an iron-tip was attached, which enabled the load to be supported when a rest break was taken. It is said that the old route through the mountains is pock-marked with holes worn into the rock by the porters' supporting sticks. The trade was an ancient one; tea was introduced into Tibet in the seventh century. In return the Tibetans provided horses and many other products including wool and the herbs used in traditional Chinese medicine.

As we left Ya'an we spotted a series of figures cast in bronze on a traffic island on the west road. Stopping and making our way across we found that it was a memorial to the Tea-Horse Trail porters and comprised a representation of men and women with their enormous loads, a fitting tribute to all those who had been engaged in the trade.

The statue of the tea porters by the road into Ya'an. Remarkably the central piece is clearly modelled from the man in Wilson's photograph (p. 43).

The famous and historic iron bridge at Luding has provided safe passage across the raging Dadu River for many generations.

Continuing, we followed an even-surfaced road into the mountains, along the course of the Tianquan River. The road climbed towards the much feared Erlang Shan, a watershed range separating the drainage basins of the Min and Dadu rivers. Today the old road over the pass has long been abandoned as the new road passes right through the mountain via a modern tunnel. It is still possible to negotiate the track along which the Tea-Horse Trail passed, though the rich vegetation for which this mountain is famous is gradually enveloping the track. Soon all vestiges of this tortuous path will pass into history and the countless generations of porters who struggled with their immense loads to reach distant Kangding will become a folk memory.

The tunnel disgorges increasing amounts of traffic into the Dadu Valley just above the town of Lengji, which could be seen on the other side of the river. It was a warm and humid afternoon as we emerged from the tunnel and it was clear that little rain had fallen recently in this essentially arid valley. The river was a chocolate-brown snake far below, swollen by snowmelt from the mountains far away to the north-west. A line of towns and villages can be found along the river valley and the road took us down to a most famous place: Luding. This has been a crossing point on the river for a very long time. In the early eighteenth century an iron suspension bridge was built to carry traffic across the river, no doubt replacing an earlier flimsy bamboo construction. Even today the bridge is an impressive structure, over 100 metres long it provides safe passage over the raging Dadu River. It has passed into legend because of a supposed action which took place here during the Long March of Mao Zhedong's Red Army. It is reported that on the 29 May 1935 a

A seasonal market in Luding with bamboo baskets brimming with locally-grown walnuts and Chinese chestnuts.

81. View on the Tung River near Lu-ting-chiao, W. Szech'uan. 4,000 feet. Showing erosion by mountain torrents. Alnus cremastogyne Burkill, in bed of stream. August 1, 1908.

'View on the Tung River near Lu-ting-chiao, W. Szech'uan. 4,000 feet. Showing erosion by mountain torrents. *Alnus cremastogyne* Burkill, in bed of stream. August 1, 1908.'

There is little perceptible change in this part of the valley from when Wilson stood on this spot, even down to the rocks in the river bed.

small contingent of Red Army troops managed to secure this bridge against murderous fire from much superior numbers of opposing Guomindang troops stationed on the western shore. This enabled the rest of the soldiers to cross the Dadu River and continue on to the security of the base that Mao had established at Yan'an, in the neighbouring province of Shaanxi. Unfortunately this incident seems to have been a pure fabrication, part of the myth that Mao built up around himself. In a devastating recent biography Jung Chang and Jon Halliday have shredded much of this myth and laid bare a new and far more sinister Mao. According to Chang and Halliday there were no Guomindang forces within 50 kilometres of Luding and therefore no battle could have occurred.[9]

On crossing the river by the adjacent concrete structure, built in the 1950s to carry heavy lorries and troop transports, we continued up the river to another photocall. Wilson's image shows a typical view of the valley, a view that our driver Mr Wang quickly recognised. Two villages are shown which he told us were Chu Ba and Shang Tian Ba, though Wilson doesn't name them. Mr Wang was able to be so precise because he had moved to Xia Tian Ba, the next village along – not visible on Wilson's image – for two years as a 16-year old in 1969, exiled from his family as part of the vast movement of people into the countryside that had occurred due to the Cultural Revolution. We were able to match the image exactly and noted how little it had changed in the intervening years. Having accomplished this we were left with the feeling that we were now really getting into our stride. One element that became immediately obvious in rural areas and was repeated time and again during the next fortnight was the fact that in trying to locate exactly where Wilson stood we needed to find the most obvious and easiest position. Wilson used a heavy whole-plate Sanderson camera and wooden tripod, so logic suggested

Tony carefully sighting the Dadu River images. Wilson knew this river as the Tung River.

The ancient town of Lengji set beside the Dadu River was visited several times by Wilson.

'*Ginkgo biloba* Linn. Tree 80 x 25 feet, and shrine. Village of Leng-che, valley of Tung River, W. Szech'uan. 3,000 feet. August 1, 1908.'

230. Ginkgo biloba Linn. Tree 80 x 25 feet, and shrine. Village of Leng-che, valley of Tung River, W. Szech'-uan. 3,000 feet. August 1, 1908.

Old trees are greatly revered in China, often being ascribed with mystical powers, and are locally venerated as a consequence. The shrine still remains at the base of this great tree.

that he was likely to seek the easiest positions possible, except where a particular image demanded more. 'Photography in the forests is no mere pastime,' he tells us. 'It took over an hour on three occasions clearing away brushwood and branches so as to admit of a clear view of the trunk of the subject.'[10] Rather than complicate things we increasingly began to ask, where is the most obvious position? Time and again this proved to be the right position.

Our next stop was at the village of Lengji, to see the famous and much admired ginkgo tree. Wilson was charged by Sargent not just to photograph landscape and habitat scenes but individual trees, particularly if they were notable for their rarity or great size. Not surprisingly, ginkgo trees, with their longevity and cultural and religious significance, featured regularly amongst his photographs. China abounds in old specimens of *Ginkgo biloba*, many of which are several thousand years old. Tony and I had visited the Lengji tree in 2001 and though we were familiar with Wilson's image of this specimen, taken on 1 August 1908, we knew that recent building work in the village makes it impossible to take a corresponding image today. Indeed it is impossible to take any sort of image of the tree due to its great size and now restricted situation. Nonetheless, we were still keen to pay our respects to this great matriarch – as it is a female tree – claimed to be at least 1,700 years old. As we stood admiring the tree a line of kindergarten children emerged from an adjacent street, all in single file and with winning smiles and giggles. With some gentle coaxing and the encouragement of their guardians we managed to persuade them to pose with Tony in front of the old tree for some charming photographs.

All too soon it was time to go and we said goodbye to the old ginkgo before rejoining the vehicles and recommencing our journey. The road deteriorated badly and we gradually lost the sunshine we'd enjoyed all day. We climbed out of the Dadu Valley west towards Hailoguo, one of the valleys that tumbles from Gongga Shan, Sichuan's highest peak. There had been some serious

Tony and the children of Lengji, dwarfed by the massive bole of the old ginkgo.

Mark and Tony at the base
of the now dead Chinese
fir, lighting incense sticks
in homage to Wilson.

rockslides and the road was being rebuilt. Eventually we arrived at Moxi, site of the old dead Chinese fir (*Cunninghamia lanceolata*) that had acted as the catalyst for our quest to follow Wilson. There was still enough daylight left to visit the deceased tree and we headed up the street towards its location. Though we had previously visited the site and undertaken a good deal of research we still learnt some new facts about the tree and its fate. A posse of wonderful old ladies – part of a roster of people who look after the attendant temple situated at the base of the tree – provided us with great photo opportunities and more information, including the location of another old tree at the next village. Tony and I lit incense sticks in memory of Wilson before taking our leave. Heavy rain meant a sprint back to the hotel and a hearty supper before a good night's sleep.

The following day we decided to investigate the tree that the temple ladies had told us about. It was located at Xin Xian, about two to three kilometres outside Moxi. This was no great diversion as it lay on a new route to Kangding that Dr Yin advised we should take as it would allow us to arrive much more quickly than the conventional route through the Dadu Valley. The tree was easy to locate, lying just above the road in a field of maize. As with so many of these specimen trees it had a small temple at its base. We quickly identified it as a species of wingnut, a member of the genus *Pterocarya*. Though its exact identity was something of a puzzle, we eventually decided, with no great conviction, that it must be *P. hupehensis*, a species native to Sichuan. It was obviously a relatively young and vigorous tree and in speaking to local people, including one old lady that joined us in the middle of eating her breakfast, it was clearly not much more than 100 years old. Still, an interesting and unusual tree, though one which has a great many years ahead of it before it can rival the Lengji or Moxi trees.

We immediately had some bad news: our shortcut north to Kangding was not possible because the bridge carrying the road had been destroyed by heavy rains. We had to retrace our journey of the previous day and go back through the Dadu Valley to Luding before turning towards Kangding. This was a tedious, though straightforward, journey along the same bumpy road. We crossed at Luding in good weather at about midday and headed up the Kangding River. This was old territory for us and for Wilson, who had visited Kangding by this route on several occasions.

The pendulous fruits of *Pterocarya hupehensis* at Xin Xian, a supposed veteran tree that we concluded was barely a centenarian.

In a land of rivers, bridges play a vital role and, judging by his photographic records and writings, Wilson was fascinated by bridges, not just out of necessity, but out of a real interest in their construction and position. The Zheduo River, to Wilson the river Lu, bisects the town of Kangding and when joined by the Yala River tumbles south as the Kangding River, falling over a thousand metres in elevation in a relatively short distance. As a consequence it is an angry, turbulent river. Wilson photographed a bridge over the river beside the road to Kangding on 30 July 1908 and his position was easily found. The bridge had long since disappeared, though several of the rocks still bore the holes that had held the supporting posts. Little else had altered; the river had lost none of its venom and was still a noisy, roaring torrent.

The characterful and helpful old lady at Xin Xian discussing the history of the village and its trees.

Despite the arrival of tourism, Kangding still remains an important market town and locally-gathered products like sprays of *Juniperus tibetica* are still widely used.

'Tachien-lu River, W. Szech'uan, view in ravine. 6,000 feet altitude. This stream falls 4,000 feet in less than 20 miles! July 30, 1908.'

72. Tachien-lu River, W. Szech'uan, view in ravine. 6,000 feet altitude. This stream falls 4,000 feet in less than 20 miles! July 30, 1908.

Though the bridge has disappeared, little has altered from the day that Wilson took his photograph.

66. Tachien-lu, W. Szech'uan, from the south. Altitude 8,400 feet. July 23, 1908.

Despite the preponderance of new buildings, Kangding is still dominated by the rugged scenery which surrounds it.

67. City of Tachien-lu, W. Szech'uan. Altitude 8,400 feet. The great trade entrepôt of the Chino-Thibetan borderland. July 23, 1908.

'City of Tachien-lu, W. Szechuan.
Altitude 8,400 feet. The great
trade entrepôt of the Chino-
Thibetan borderland.
July 23, 1908.'

Finding the spot where Wilson took this photograph was extremely difficult, as the town of Kangding (Tachienlu) had spread into the surrounding foothills.

We entered Kangding in the early afternoon, allowing us plenty of time to search out images Wilson took of the town of Tachienlu, as it was known to him. Once again Mr Wang proved to be invaluable. He had lived in the town for many years during the time that he drove transport lorries over the high mountain passes to Lhasa. He took us to what looked like very unpromising spots but once again he was dead right. Although there were lots of new buildings the positions were spot on. Wilson writes: 'The town is hemmed in on all sides by steep, treeless mountains whose grassy slopes and bare cliffs lead up to peaks culminating in eternal snow. On the whole, the situation is about the last in the world in which one would expect to find a thriving trade entrepôt.'[11] These statements remain true today, though tourism is now rivalling the income gained from trade and manufacture. With Mr Wang's help we accurately matched the images showing the steep slopes around the town. A third image was another of more than passing interest. Like many travellers in the East Wilson was much taken by the complex relationship between the Han Chinese and the Tibetan people. He studied the latter in great detail, their tribal arrangements, manners,

70. View of Lamaseries outside Tachien-lu, W. Szech'uan. 8,500 feet. The farther belongs to the "yellow" sect; and the nearer and principal to the "red" sect. July 27, 1908.

'View of Lamaseries outside Tachien-lu, W. Szechuan. 8,500 feet. The farther belongs to the "yellow" sect; and the nearer and principal to the "red" sect. July 27, 1908.'

Perhaps the key difference here since 1908, apart from the additional buildings, is the abandonment of the cultivated fields and the gradual encroachment of native vegetation.

customs and their religion. His image shows two lamaseries on the outskirts of Kangding, from the two differing sects of the religion, yellow and red. The same view today showed that the monasteries have been much embellished but remain essentially the same, though many new buildings have sprung up around them in the intervening years.

A very productive few days had been achieved. Not only had we precisely matched several of Wilson's images but we had discovered important aspects about how he went about his work. The fact that common sense dictated that he seek easy locations to set up his cumbersome equipment and that he worked with speed and precision would be important in the more arduous work we needed to accomplish in the mountains. We had also begun to feel a real sense of the man and his time and had proved beyond doubt that his passing was not so remote from us. We started to look forward to an exciting adventure in the mountains above Kangding: the Da Xue Shan (Big Snow Mountains).

CHAPTER THREE

MYSTERY TOWERS OF DANBA

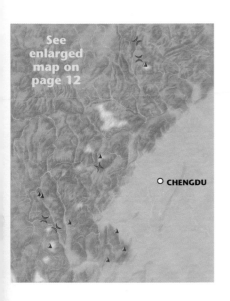

See enlarged map on page 12

○ CHENGDU

Tony and Mark take a rest from the biting cold winds amongst the prayer flags on the Ya-jia Pass at 4,000 metres elevation.

Tatien-lu is a small and filthy dirty place, it boasts a large mixed population of Chinese and Tibetans. Being on the highway from Pekin to Lhasa, officials are constantly passing and re-passing. This makes it a highly important place, both politically and commercially. Although Batang, 18 days journey to the west, is the actual frontier town, Tatien-lu is really the gate of Tibet.[1]

Wilson's accurate but rather unflattering description of Kangding was penned at the conclusion of his first visit to the town in 1903. Wilson had made the journey to Kangding on the instructions of the Veitch nursery who wished to add a very special plant, the lampshade poppy (*Meconopsis integrifolia*), to their nursery catalogue. 'Messrs Veitch despatched me on this second, and very costly, journey to the Tibetan border for the sole purpose of discovering and introducing this, the most gorgeous alpine plant extant,' recorded Wilson.[2]

Kangding still has a frontier town feel about it and is inalienably a Tibetan place. It also remains a very important staging post on the road that leads westward into Tibet. This road was one of the great highways constructed during imperial times to hold the Celestial Empire together. But Chinese writ did not extend very far. Indeed the country around Kangding remained lawless and untamed until very recent times. Historically the area was known as Kham and its inhabitants, the Khampa, were much feared for their ferocity and war-like demeanour. In truth, Khampa was a collective noun as the area was home to a very diverse group of related, though distinct, peoples. For a thousand years, after the collapse of the vast Tibetan Empire in the ninth century, Kham remained unconquered and unconquerable, its peoples engaged in ceaseless internecine conflict as petty warrior chieftains battled for supremacy. Banditry was an accepted means of acquiring wealth and position.

The various groups of ethnic peoples he encountered fascinated Wilson and he wrote about them extensively. His understanding was gained both at first hand and through studying the work of contemporary ethnographers, though this understanding was far from exact. For example, he used the Chinese generic, and derogatory, name 'Sifan' (western barbarian) to describe the tribe now identified as the Qiang, one of the 56 official ethnic peoples in China. To the enquiring Edwardian mind the alien culture and manners

Meconopsis integrifolia subsp. *integrifolia*: the principal objective of Wilson's second trip to China was to introduce this plant to western gardens.

The Khampa people still retain many traditions including distinctive apparel and the widespread use of locally bred horses.

of the various peoples, particularly their peculiar (and supposedly immoral) sexual liaisons in which both polyandry and polygamy were commonplace, was of abiding interest. Their relative lack of sophistication also appealed to Wilson, suggesting to him a oneness with their environment that he found endearing.

The eventual subjugation of the Kham region in what the Chinese called the 'peaceful liberation of Tibet' finally ended the brigandry and general lawlessness in the 1950s. Eastern Kham formally became part of Sichuan Province and western Kham formed a large part of the Xizang Autonomous Region. Despite this, the Khampa retain their individuality by virtue of their strong culture and association with their land. In travelling this country it is impossible not to be impressed by their proud bearing and independent mien and it is easy to understand how they struck fear into the hearts of friend and foe alike. Despite the suppression of banditry, in recent years occasional acts of violence against foreigners still occur when travellers are held-up by groups of armed local men: old habits die hard.[3] Through the 1980s and 1990s the Kew expeditions to this part of China employed the services of an armed Chinese policeman, Lao Liu, as a precaution against unwanted local attention, though he never drew a weapon in anger!

Wilson's quest for the lampshade poppy was, therefore, into territory that he knew little of and amongst people with whom, at that stage, he was largely unfamiliar. Not only was the territory unfamiliar, it was built on a grand scale. The Da Xue Mountains, into which Wilson was travelling, together with the neighbouring ranges form part of the vast, complex Hengduan Shan, the eastern extension of the Himalaya. They were created at the same time but, due to the shearing effect involved when the landmass of India collided with the Asian continent, they incline north/south. These mountains, eroded by monsoon-swollen rivers – the Jinsha, Yalong, Dadu and Min – form an enormous convoluted mass of peaks, ridges and spurs with deep, sheer-sided valleys. The range climaxes at the summit of the mighty Gongga Shan, which at 7,556 metres is Sichuan's

A view to distant Gongga Shan, showing its distinctive peak.

Photograph: W. McNamara, Quarryhill Botanical Garden.

highest mountain by some way. Joseph Rock brought Gongga Shan to the attention of the West in 1930, when he infamously over-estimated its height, erroneously claiming it to be higher than Mount Everest.[4] Wilson would be travelling at far higher elevations and over much more demanding terrain than he had experienced in his first trip to the more modest hills and valleys of Hubei.

He was not alone, however. It is intriguing that Wilson rarely mentions any western companions in his writings, let alone provides any details of their backgrounds and occupations. This time, as he prepared to find the lampshade poppy, he was accompanied by an experienced traveller. On July 16 1903 he started out for the mountains: 'On this journey I was accompanied by Mr Edgar of the China Inland Mission, in whom I found a delightful companion . . . Leaving by the South Gate we followed the main road to Lhasa – a broad, well-paved road.'[5]

Fortunately, we can both reveal and understand Mr Edgar, a person of considerable interest. The Reverend James Huston Edgar was a missionary in the employ of one of the most unusual of the Christian organisations that proselytised throughout China. The China Inland Mission was founded in Brighton, England on 25 June 1865. Its founder, J. Hudson Taylor, took a radically different view of evangelisation. His organisation was non-sectarian and preached a message based on faith and prayer. He recruited from across the social spectrum and many of his most fervent acolytes were members of England's working class. Women also formed a significant number of the China Inland Mission both as wives and as missionaries in their own right. Members of the China Inland Mission entered China in 1866 in the wake of the Treaty of Tientsin, which guaranteed the freedom of missionary activity.[6] They adopted Chinese dress and customs and attempted to integrate with the local populations as much as circumstances would allow. By the time Wilson came on the scene the China Inland Mission had stations throughout China and was especially active in the more remote western provinces. They suffered disproportionately during the Boxer Rebellion, when missionaries above all other western people were the target of anti-foreign violence aroused by this populist movement. In the northern province of Shanxi during June, July and August 1900, the China Inland Mission station was slaughtered – men, women and children – with executions taking place in the most horrific way imaginable. J. H. Edgar, who escaped the worst of the Boxer outrages, spent all his adult life in China, and became a noted sinologist and scholar of the local Tibetan tribes.[7] At the end of a life marked by self-sacrifice and privation he died at Kangding on 23 March 1936 at the age of sixty-four, continuing right up to the end with the work that had been his vocation. Wilson was in good hands.

The main road to Lhasa was a well-travelled highway but not one that Wilson would remain on for long. It quickly rises to the Zheduo Pass, which today is still the most commonly taken route into Tibet from western Sichuan. On the flanks just below this pass the lampshade poppy can be easily found and many writers have assumed that this is

From the Zheduo Pass the road back to Kangding snakes its way through the rugged peaks of the Da Xue Shan.

where Wilson gathered his first plants. But the Zheduo Pass was not Wilson's destination; he was heading for the Ya-jia Pass, which followed an alternative and much less-used track to the south.

At first the journey was enjoyable and Wilson revelled in his surroundings: 'our road was through lovely grassy country, with a steady rise. A wealth of many coloured herbs enlivened our path,' and, 'we continued through similar country, with a fine snow-clad peak straight in front of us and another to our left'. Soon, however, the going became much tougher, and heavy rain fell as they reconnoitred the mountainsides close to their overnight stopping point. The altitude had a detrimental effect on his coolies and all endured a miserable night.

Our journey up to the Ya-jia Pass was rather more comfortable. The road was well-surfaced right to the top, though the occasional small landslip had to be carefully negotiated by the vehicles. As Wilson suggested, snow-clad peaks were visible all around and we were fortunate to have fine weather in which to appreciate them. Looking back, a stunning range of mountains could be seen to the north-east beyond Kangding – the Lian hua Shan (Lotus Flower Mountains) – no doubt one of the views enjoyed by Wilson during his own ascent 103 years before. Ahead the scene was much less promising with dark clouds scudding across the sky alternately revealing and concealing the mountain tops and providing tantalising glimpses of the pass.

Frequent small landslips can temporarily make these high mountain roads impassable, necessitating some rapid excavation work.

The stunning panorama of the Lian hua Shan, a view appreciated by Wilson during his ascent to the Ya-jia Pass in 1904.

Wilson's miserable end to the day was compounded during the night:

Having at length got rid of our soaked garments – a difficult enough task under the circumstances – we eventually got between the blankets. No sooner had I lay down than a drip came a spot of rain into my eye: I turned over and drip came another into my ear. I twisted this way and that way, but there was no escape. Like evil genii these rain-drops pursued me turn which way I would. I could not move my bed, since this was longer than the tent was broad, and my feet already exposed, and we sorely afraid the whole thing might collapse, it being anything but secure . . . About 4am our firewood gave out and things assumed a very dismal aspect. However, all things have an end; day at length dawned and all were devoutly thankful . . . With what fire remained we managed to boil some water and make some tea. We breakfasted on ship's biscuits and cheese and felt none the worse for the night's experience.[8]

Wilson was, above all things, a fatalist.

The rain stopped and Wilson and Edgar prepared for the day's work, during which they hoped to find the lampshade poppy. A farmhouse, one thousand feet below their overnight position, was commandeered and their cook, who was suffering from severe altitude sickness, was taken down to recuperate. The journey to the pass began at 7am

The flora of the moist ground below the Ya-jia Pass is replete with a wonderful array of colourful flowers – (*clockwise from left*) *Podophyllum hexandrum, Rheum alexandrae, Primula secundiflora, Primula sikkimensis* and *Iris chrysographes*.

The lampshade poppy growing
close to the site where Wilson
first encountered it.

and after some initial rain showers continued on in sunshine. The alpine flowers captivated Wilson; in early summer these Chinese mountains are amongst nature's most exquisite natural gardens. Tony and I arrived at the peak of the display and we left the vehicles three or four hundred metres below the pass and proceeded on foot. By the roadside, a braided mountain stream provided ample moisture and it was in this sodden turf that the greatest diversity could be found. 'I wish I had the ability to describe this floral paradise with all its glories, but this is beyond me,' wrote Wilson.[9] I certainly won't try where Wilson failed and hope that the images reproduced on these pages will give the reader a hint of the individual and collective beauty of these mountain flowers, many of which have become firm garden favourites amongst discerning growers.

We followed Wilson's and Edgar's route to the pass knowing that at any time the lampshade poppy would appear. Our experience was almost exactly as theirs had been:

At 11,000 feet I came across the first plant of Meconopsis integrifolia*! It was growing amongst scrub and was past flowering. I am not going to attempt to record the feelings which possessed me on first beholding the object of my quest to these wild regions . . . I had travelled some 13,000 miles in five and a half months and to be successful in attaining this first part of my mission in such a short time was a significant reward for all the difficulties and hardships experienced en route.*[10]

The lampshade poppy is a monocarpic species, dying after flowering, but it produces ample seed and has proved to be relatively amenable in cultivation, particularly in the cool summer climate of Scotland.[11] Wilson's plant became an instant success, flowering in its first season in the Veitch nursery and persisting for many years.[12] All the recent trips to this and neighbouring parts of China have reinforced its presence in cultivation and it is not unusual to see this plant flowering in northern gardens. In cultivation it has also produced several attractive hybrids with other Asiatic species such as *Mecanopsis* ×*beamishii* (M. *integrifolia* × M. *grandis*) and M. × *finlayorum* (M. *integrifolia* × M. *quintuplinervia*). In recent years botanical opinion, particularly that of Dr Chris Grey-Wilson, has suggested that this variable plant is easily divisible into two distinct entities – M. *integrifolia* and M. *pseudointegrifolia,* the latter a plant with nodding and more open flowers, quite distinct from Wilson's plants that have globular and more upright flowers. [13]

After this first plant the mountainsides began to reveal a veritable cornucopia of poppies. Wilson recorded that 'as we continued the ascent, *Meconopsis integrifolia* became more and more abundant. At 12,000 feet and upwards, miles and miles of the alpine meadows were covered with this plant, but only a few late flowers remained'.[14] Being a month earlier Tony and I caught every plant in full flower, the sun-disk blooms swaying in the mountain breeze, flaunting their wares for any passing bees. Our climb continued in deteriorating conditions until we reached the pass at nearly 4,000 metres. Wilson tells us of the fear that the Ya-jia Pass engendered amongst his Chinese followers who were not, by nature or inclination, mountain people: 'this Ya-kia pass enjoys an unenviable reputation, and is much dreaded on account of its asphyxiating winds. It is said to be the only pass in the neighbourhood which "stops peoples' breath."'[15] On reaching the pass we were forced to concur, for though it was June 17, the temperature hovered around freezing point and a biting wind blew from the bleak Tibetan Plateau to the west, bringing pulses of sleet in its wake. Despite this we were thrilled to take an image at almost exactly the same location as Wilson had when he re-visited the pass on 19 July 1908.

Wilson stayed a second night on the mountain, this time in the more salubrious surroundings of the farmhouse, which had been commandeered for the use of his party. This proved to be a clean, dry, cosy dwelling, and to add further to their good fortune his cook was quite recovered and prepared a hot meal for the team.

On the next day of our trip a most intriguing incident occurred, something that caught me quite by surprise. I have already mentioned Gongga Shan, the giant peak that dominates the Da Xue Shan range. It is the highest mountain in Asia outside of the main Himalayan chain and it exerts a baleful influence. Numerous glaciers grind their way down its flanks and such is its size that it generates its own climatic conditions over a sizeable swathe of the surrounding country. This mountain has always fascinated me. Despite dominating the area it is frequently covered in cloud: I have journeyed to five key vantage points – east, north-east, west and south of the peak – and been disappointed to find a shroud-covered summit every time. In all his writing Wilson never mentions this mountain either as Gongga Shan or Minya Konka, its Tibetan name. How can this be?

65. Tachien-lu, near, W. Szech'uan. The Ya-chia-k'an snows and alpine regions clothed with dwarf junipers and rhododendrons. 13,000 feet. July 19, 1908.

'Tachien-lu, near, W. Szechu'an. The Ya-chia-k'an snows and alpine regions clothed with dwarf junipers and rhododendrons. 13,000 feet. July 19, 1908.'

The barren and desolate Ya-jia
Pass in 2006, unchanged since
Wilson and Edgar first came here.

During his visits to China he spent many months in the Da Xue Shan, surely he must have heard some local reference to the peak or glimpsed some distant view? Given his silence on the matter the obvious conclusion was that he was also unlucky and never had a clear view of the summit nor did he hear mention of it amongst the local people.

As Tony and I wandered the lonely slopes around the Ya-jia Pass I pondered this matter, knowing that the giant mountain lay to the south-west of our position. All around us were shattered and snow-clad peaks. It would have taken a strenuous hike into the higher reaches to breast these in order to provide an unencumbered view to the south-west, and time didn't allow this opportunity. In the warmth and comfort of our 4 × 4 as we took the road back to Kangding, I re-read Wilson's account of his first journey to the Ya-jia Pass, particularly the second day of his visit. One paragraph leapt from the page. Although I had pored over all Wilson's writings for the best part of the previous 18 months, the significance of the words had, until now, escaped me:

The moraine in front of us terminated in tremendous fields of ice, glaciers of a virgin peak, 21,000 feet high. The sun shone brilliantly and we got a magnificent view of the surrounding mountains. South, south-west of us lay a gigantic peak, several thousand feet higher than the one mentioned; its summit crowned with snowfields of enormous size.[16]

Gongga Shan? Surely.

The following day it was time to move on to the next phase of our journey. Wilson's first trip in the employ of the Arnold Arboretum, his third visit to China, took place between 1907 and 1909. Released from the economic shackles imposed by the Veitch Nursery he could take a much more expansive view of his activities. The patrician director of the Arnold Arboretum, Charles Sprague Sargent, encouraged Wilson not only to

Mark at the Ya-jia Pass – this location has long enjoyed a malevolent reputation because of its fickle weather conditions and rarefied air.

The hardy and ubiquitous *Rhododendron przewalskii* covers huge areas of the high mountains above Kangding.

'science-up' his work – more emphasis on herbarium specimens and greater attention to field notes – he also insisted that a comprehensive photographic record of the journeys be produced. In a letter to Wilson dated 6 November 1906, a copy of which can be found in the Wilson archive at the Arnold Arboretum, Sargent explains:

> *I write to remind you of the very great importance of the photographic business in your new journey. A good set of photographs are really about as important as anything you can bring back with you. I hope therefore you will not fail to provide yourself with the very best possible instrument irrespective of cost.*

Sargent's prescience not only provided us with an excellent series of images of plants and landscapes, which were later published by the Arnold Arboretum, but also a snapshot of Imperial China right at the end of its long history; within a year of Wilson's departure China was effectively a republic.

Thus equipped and instructed Wilson arrived at Yichang, his old base on the Yangtze River, in February 1907 for what was to be his most successful trip, a trip that cemented his reputation as the foremost collector of his generation. I have long felt that the second year of this expedition, 1908, was also his most interesting and productive and in following in Wilson's footsteps I was especially keen to emulate some of his travels during that year. From Kangding we had the opportunity to retrace Wilson's journey of June–August 1908 when he travelled between Dujiangyan (Kuan Hsien) and Kangding, though we would travel it in reverse. Interestingly, Wilson himself was following an earlier traveller – Sir Alexander Hosie – as he tells us:

> *During the summer of 1908, when in Chengtu, I determined upon a journey to Tachien-lu.*
> *Previously, in 1903 and again in 1904, I had visited this town by three different routes.*
> *This time I decided upon following the road leading from Kuan Hsien via Monkong Ting*
> *and Romi Chango. The only published account of this route that I had knowledge of is a*
> *report by Mr (now Sir) Alexander Hosie, erstwhile HBM's Consul-General at Chengtu, who*
> *returned over this road in October 1904.*[17]

The account Wilson refers to, 'Journey to the Eastern Frontier of Thibet', was published as a Parliamentary Report and presented to Parliament in 1905. Hosie took the same direction as Tony and I would, east from Kangding, which he left on 10 October 1904. This was by no means a regular or accepted highway and that is what interested Wilson: 'what I saw of the forests and mountain scenery, together with the quantity and variety of the plants discovered and collected, abundantly repaid me for the hardships experienced'.[18] My hope was that we could also experience some of this scenery and plant diversity. But could we retrace the route and match some of the many images that Wilson had taken on this journey?

Things began disappointingly. Wilson had travelled on foot on the east side of the Da Xue Shan between Kangding and the village of Hsin-tientsze and even today there is no suitable road for motorised vehicles. This meant that we would have to drive up the west side of the range before rejoining Wilson's route beyond Hsin-tientsze. Fortunately, apart

Prayer flags on the Zheduo Pass. The Tibetan word for prayer flag is Dar Cho. 'Dar' means to increase life, fortune, health and wealth. 'Cho' means all sentient beings. Colourful prayer flags fluttering in the wind are simple devices that, coupled with the natural energy of the wind, quietly harmonise the environment, impartially increasing happiness and good fortune among all living beings. They are not just pretty pieces of coloured cloth with strange writing on them. The ancient Buddhist prayers, mantras and powerful symbols displayed on them produce a spiritual vibration that is activated and carried by the wind across the countryside. The tradition has a long continuous history dating back to ancient Tibet, China, Persia and India. The meanings behind prayer flag texts and symbols, indeed behind the whole idea of prayer flags, are based on the most profound concepts of Tibetan Buddhist philosophy.

from stunning views of some of the snow-clad peaks and a range of hot springs at Je-shui-t'ang, it seemed we would miss nothing of great import. No images of particular interest record this section of the journey. We left Kangding taking the road up and over the Zheduo Pass. In the sunlight the roadsides were bright with wild flowers, many of a striking nature, including the large flowered but short-statured Tibetan lady's slipper orchid (*Cypripedium tibeticum*) with large maroon pouches. At the pass we had something of a shock. Having been at this lonely spot in 2001 we were dismayed to find that things were much changed. A wooden belvedere had been built about 150 m above the pass, reached by a flight of steps, and another building was under construction nearby. No doubt these developments are underpinned by good intentions, this spot is very much on the tourist route, but it somehow seems quite inappropriate to despoil these pristine alpine areas with such frippery. We didn't dwell.

On the other side of the pass we got stuck behind an endless convoy of lumbering army trucks, which slowed the pace considerably. One of the positive results of this inconvenience was that we were able to continue to admire the carpet of flowers in the grassy alpine pastures. Unlike the valleys to the east, the west side of the Da Xue Shan is quite dry and few trees are to be found. As a result extensive grasslands are a feature and at this time of year they boast a display worthy of the most colourful flower garden. The turf was studded with gorgeous plants – *Incarvillea compacta* var. *qinghaensis*, *Meconopsis racemosa*, *Lilium lophophorum* – odd specimens of *Rhododendron capitatum* formed hummocky mounds amongst the grass sward and the horizon was an endless, undulating green line. Eventually we turned north leaving the army to continue their

Cypripedium tibeticum growing in yak-grazed meadow on the edge of spruce forest.

Flowers of the high passes and grasslands of the Da Xue Shan – (*from left*) *Lilium lophophorum,* *Incarvillea compacta* var. *qinghaensis* and *Meconopis racemosa.*

The local flora is occasionally put to a novel use; this Tibetan girl is using incarvillea as a kazoo!

procession into Tibet. The road became more and more potholed and uneven as we proceeded. Along the way the solid architecture of Tibet began to dominate, with farm buildings of substantial size and construction. Many are only seasonally occupied as the inhabitants leave in the spring to spend the summers in the high mountain pastures grazing their herds of yak. We passed through the important religious centre of Tagong, dominated by its richly decorated and ornamented temple. Rising in front of us was another range of impressive peaks, the Da Pao Shan (Big Cannon Mountains). This linked us back with Wilson, who enjoyed fine weather during the last leg of his journey, albeit on the other side of the range to our position:

> *the view from the summit of the pass far surpassed my wildest dreams. It greatly exceeded anything of its kind that I have seen and would require a far abler pen than mine to describe it adequately. Straight before us, but a little to the right of our view point was an enormous mass of dazzling eternal snow, supposed to be, and I can well believe it, over 22,000 feet high. Beneath the snow and attendant glaciers was a sinister-looking mass of boulders and screes.*[19]

Unfortunately for us low clouds obscured the actual summit, though Xiao Zhong told us of that on a recent previous visit he had seen nothing of the mountains at all, so perhaps we were not so unlucky.

Typical solidly-built Tibetan houses on the road to Tagong.

The new temple at Tagong with the massive bulwark of the Da Pao Shan behind.

The summit of Big Cannon Mountain remained obscured by clouds for most of the day.

The valley of the Mao Niu River is an extraordinary place with pristine forests of spruces, pines, birches and many other broadleaved trees.

Gradually more and more trees began to reappear as we descended into a river valley and passed through the mountains to the eastern side. Soon the river was a raging, boiling torrent hemmed in by near vertical cliffs rising perhaps 1,000 m upwards. At some point we struck Wilson's route from the north, though it was difficult to be exact. We looked for the village of Kuei-yung, where Wilson stayed on the night of 6 July 1908 and enjoyed an unusual experience:

This place consisted of half a dozen houses, purely Tibetan in character, built on a slope and surrounded by a considerable area under wheat, barley and oats. The mountains all around are heavily forested with coniferous trees and in the far distance a snow-capped peak glittered on the horizon . . . The house we lodged in is three-storied with the usual flat mud roof . . .The housewife, a most cheery if dirty person, had a very musical laugh. Things generally appeared a joke to her, and incited her to frequent laughter, which it was pleasant to hear. My followers were oddly amused at the strangeness of things and appeared to enjoy the novelty.[20]

Try as we might we could find no sign or reference to Kuei-yung and so we continued on up the river valley. Eventually we picked up the trail. The village of Mao Niu (Yak Village) was exactly as Wilson described it: 'a fair sized village for the country and mainly perched on a flat about 200 feet above the torrent and surrounded by a considerable area under wheat – a veritable oasis, in fact, surrounded by high mountains'.[21] We entered the village and caused something of a commotion, as usual the children chased along behind our vehicles with shouts and whoops. As we stopped most of the adult population emerged. Earnest discussions ensued which were more confusing than helpful. Kuei-yung, we were told, was up an adjacent side valley, not at all where Wilson placed it. The headman also related a village story, passed down through several generations, which tells of the visit of a foreigner interested in plants – probably an apocryphy for our benefit, but intriguing nonetheless. We left Mao Nui somewhat perplexed but certainly back on Wilson's exact route. The road continued to skirt the river, hemmed within its tight, vertically-sided valley, 'a narrow, savage, magnificently wooded ravine'. The road had a good level macadam surface allowing for rapid transport, very different from Wilson's day: 'my head coolie declared it was the worst road we had ever traversed, and I was inclined to agree with him'. [22] The villages mentioned in Wilson's text began to appear in quick succession, and on reaching Tung-ku, we had an image to match.

Enlisting the help of village people, old and young, at the village of Tung-ku.

Wilson's photograph shows a tall cypress tree against a mountainous background with dwellings in the middle distance. Things had changed markedly in Tung-ku and even with the help of some old residents we found it hard to locate Wilson's position. Tony ended up climbing onto the roof of a building, through a trap door, to get anywhere near the same image. The cypress tree had gone, a victim, apparently, of the installation of electricity into the village in the 1970s when many trees were felled to give access for the electricity cables.

195. Cupressus torulosa D. Don. Tree 80 x 10 feet. West of Romi-chango, W. Szech'uan. 8,000 feet. July 2, 1908.

'*Cupressus torulosa* D. Donn. Tree 80 x 10 feet. West of Romi-chango, W. Szech'uan. 8,000 feet. July 2, 1908.'

Despite the loss of the cypress trees, Wilson's image could still be matched by using the background hillsides.

The important town of Danba has changed greatly since Wilson's time. The Dadu River, a confluence of the Big and Little Gold rivers, begins its journey south from here.

Similar problems were also experienced at Danba, Wilson's Romi Chango, as we emerged out of the river valley. Danba is now a large, built-up town. Extensive construction made it difficult to locate where Wilson had taken his image of the town and we concluded that the site had probably been built over. Wilson was less than impressed with Danba, though he was treated with respect and courtesy:

Romi Chango, or Chango, as it is commonly called, is a poor, unwalled straggling town of about 130 houses . . . It is built on the right bank of the Tachin Ho, at a point where the river making a right-angled turn to the northward is joined by a very considerable torrent from the west. The Tachin, a river 100 yards broad, with a steady current and muddy water sweeps around majestically. High cliffs on the left bank, steep mountainsides on the right, lofty mountains to the east and west wall in the town, at the western end to which a massive square tower stands sentinel . . . In Chango we lodged at a comfortable inn, having a clean room, well removed from the street and overlooking the river. We spent a quiet day resting and refitting for the final stage of our journey to Tachien-lu. The people were not over-inquisitive and those in charge of the inn were exceedingly obliging.[23]

55. Romi-chango. View west, showing cliffs conifers and general vegetation. W. Szech'uan. 8,750 feet. July 3, 1908.

'Romi-chango. View west, showing cliffs conifers and general vegetation. W. Szech'uan. 8,750 feet. July 3, 1908.' An image that we struggled to locate, due to the major construction that has occurred since Wilson's visit.

Our experience was remarkably similar. We lodged at a comfortable hotel with a quaintly English name – Old Castle Hotel – and in the evening we enjoyed the company of local tourist officials over a very delicious meal with lots of spicy, potent dishes.

Most importantly, our hosts were able to provide us with significant local details some of which seemed to have escaped Wilson. The river we had followed along the valley was the Mao Niu River, which joined the much larger Da Jin Ho (Big Gold River) right outside the Old Castle Hotel. Further along, at the edge of Danba, it was joined by the Xiao Jin Ho (Little Gold River) and left the town as the Dadu River, heading due south to join the

The girls of Danba are widely admired for their beauty.

Min River at Leshan. We were also told that Romi Chango was a local Tibetan name that means 'town on the cliff': quite apt. The fact that the old name had fallen into disuse detracted not one iota from the fact that minority Tibetan groups still populate the area. In the remote mountain villages all around Danba are a diverse range of ethnic groups each with their own dialects and customs. A characteristic of the area is the presence of many tall, generally square towers, some of which are up to 30 m in height. Towers are a feature of the vernacular architecture throughout much of western Sichuan but nowhere are they as numerous as in the vicinity of Danba. Surprisingly, Wilson makes only occasional reference to these and never with any particular comment attached. In the surrounding countryside these towers are so plentiful that they can dominate the hillsides outside of the main river valleys. Such an omission is surprising given Wilson's interest in local architecture and the customs and manners of the attendant peoples. The only conclusion is that, travelling so speedily, Wilson missed most of them and was largely unaware of their great abundance and local importance.

A cluster of stone towers outside Danba – defensive watchtower or symbolic structure?
Photograph © Yin Kaipu

In recent years these enigmatic towers have generated a great deal of interest: who built them, when were they built, why were they built? A vivacious French anthropologist, Frederique Darragon, became captivated by these towers and has made an intensive study of them. Darragon has travelled throughout western Sichuan and eastern Tibet since 1998, documenting these 'watchtowers', as she styled them. Her records now include over 150 structures and amongst other research projects she has dated several towers to be older than 1,200 years. Darragon's work culminated in a critically-acclaimed documentary 'Secret Towers of the Himalayas', aired on the Discovery Channel in 2003. With the proceeds from this she set up the Unicorn Foundation, not only to conserve the towers but also to promote educational opportunities for the children of the local ethnic communities. Her research into the origins and purpose of these towers has been inconclusive. She believes they may have religious and symbolic significance but many other scholars take a more prosaic view, seeing them as signal stations or defensive structures. Certainly the towers are formidable edifices. When the turbulent past of these mountain valleys is considered this would seem to be the most logical explanation.

Wilson clearly knew about the violent history of the areas he was travelling in; he talks with authority about the subjugation of the mountain tribes by the Chinese in the eighteenth century and misses none of the key facts. The valleys that Tony and I traversed saw ferocious fighting as the greatest of the Qing (Manchu) emperors, Qianlong (1711–1799), sought to extend and consolidate the Chinese empire and its spheres of influence. This he accomplished by his Ten Great Campaigns, which saw him enlarge the area under Qing control in central Asia by defeating the Dzungars in Xinjiang, suppressing a rebel movement in Taiwan, and browbeating all the kingdoms on his southern borders. To the south-west, the mountain tribes on the borderlands of Sichuan and Tibet had caused considerable problems for the previous Ming Emperors and Qianlong was determined to crush them once and for all. However, despite the might of his armies – in the second campaign it was claimed that he put 200,000 troops in the field – this proved remarkably difficult to achieve. Indeed, the battles against the Jinchuan tribes, as they were described at the time, cost the Chinese treasury more money and the Chinese army more men than all the other battles put together. This was a consequence of the extreme topography that the battles were fought within, the doggedness of the rearguard action of the tribes and their defensive towers. A succession of Chinese generals were defeated and returned in disgrace to Beijing and a ritual beheading before the celebrated Agui, one of Qianlong's most trusted advisors, who brought the tribes to heel with the use of a western cannon and a wily strategy that eventually pitted the tribes against each other. Thereafter the Manchu armies wreaked a terrible revenge on the tribes for their resistance. The local king, Sonom, was taken in chains to Beijing where, according to Wilson, 'after a grand court ceremony he was sliced to pieces'.[24] Modern estimates suggest that 80 per cent of the local population were killed or displaced in the bloody aftermath of the conquest. Today the descendants of these brave fighters are still to be found in the river valleys north and east of Danba and are known as the Ergong or Gyarong, though their unique identity is subsumed within the official categorisation that places them as part of the more generic Tibetan ethnic minority.

The following day, as we left Danba and followed Wilson's route along the Xiao Jin Ho, I pondered these traumatic events. In the sunshine it was difficult to imagine the barbarity of the fighting which occurred in the valley, all was serene and the tumbling river kept its secrets to itself.

The Emperor Qianlong reasserted China's hegemony in the west. One of the greatest of the Qing emperors, under whose vigorous rule China became strong and prosperous.
Photograph © RMN/ © Thierry Olliver

CHAPTER FOUR

THE DREADED
PAN-LAN SHAN

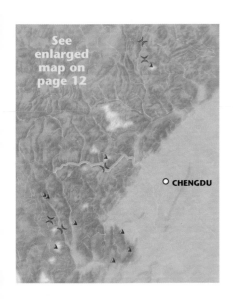

See
enlarged
map on
page 12

○ CHENGDU

In the upper reaches the Xiao Jin
Ho is a frenzied and turbulent river
little altered from Wilson's time.

Our journey up the Xiao Jin Ho would take us over 100 km to its source below the Balang Shan Pass and directly along Wilson's route of June and July 1908; it promised to be of great interest and excitement. The river was initially loud and frenzied, in Wilson's words 'one long succession of cataracts and strong rapids, the turbulent waters being thick with brown mud,'[1] but given that we travelled up-stream it became progressively smaller and meeker save for the odd stretch of white water. The river valley was exactly as Wilson described it and each landmark confirmed that we were on his precise route:

> High bare cliffs predominate, but here and there occur more or less fan shaped areas under cultivation, with houses shaded by Poplar, Willow and Walnut trees . . . Maize is evidently the chief summer crop in these regions but wheat is grown, a red, beardless variety with stout ears and harvesting was in progress.[2]

This could describe the Xiao Jin Ho Valley today with one obvious exception; apple orchards have been planted in some quantity and were a feature of the day's journey.

Fortunately, Wilson's narrative allowed us to proceed with confidence despite the fact that the village names along the route appeared to have subtly changed: Wilson's Yo-tsa is today Yue-za, Pan Ku Chiao is Ban Gu Qiao and Hsao Kuan Chai is Xiao Guan Zhai. Such changes have bedevilled studies of Wilson's journeys for many years and are largely the consequence of a change in the way that Mandarin is transliterated into English. The story of how scholars mastered the translation of Mandarin in both written and vocal expressions might at first seem an anodyne subject; nothing could be further from the truth. As in so many fields of human endeavour petty jealousies, rivalry and downright bad blood have been features of this academic endeavour.

The early Chinese scholars were Jesuit priests who inveigled their way into the late Ming and early Qing imperial courts, rapidly making themselves indispensable to the Emperor and his advisors. They brought map-making skills, knowledge of horology and, incredibly, expertise in ordnance and munitions, all much valued at the Chinese court. Most were French, though Spanish, Portuguese and Italian Jesuits were also involved. As a result there was very little transliteration of Mandarin into English in the early years. Only when the Chinese begrudgingly established diplomatic links with Great Britain, after the Macartney embassy (1792–94) to the Emperor Qianlong's court,

19. Hsao-chin Ho. View in valley, with house, cliffs and Populus. Near Mon-kong Ting, W. Szech'uan. 7,800 feet. June 29, 1908.

'Hsao-chin Ho. View in valley, with house, cliffs and *Populus*. Near Mon-kong Ting, W. Szech'uan. 7,800 feet. June 29, 1908.'

Despite the loss of the central tree the view up the river valley is easy to compare. It may well be that the remaining trees on the river side are the same as those in Wilson's photograph. Most telling is the presence of maize in both images – a remarkable continuity over a period of 98 years.

did Chinese scholarship commence with English speakers. One of the first was the remarkable 11-year-old George Staunton, son of George senior, the Secretary of Macartney's embassy, who accompanied his father as the Ambassador's page. Staunton eventually became a supercargo, an officer responsible for the care and selling of the cargoes of merchant ships, with the East India Company based at Canton. Though he could speak and write Chinese fluently and was widely recognised as one of the leading scholars of his day, he did little to establish the system for the transliteration of Mandarin into English that was needed to conduct formal international affairs. This role was taken up by a series of British consular officials, most notably Sir Thomas Francis Wade (1818–1895).

Wade was born in London, the son of an officer of the Black Watch. Inevitably, he took the colours in 1841. He was posted to Hong Kong in 1842 and was involved in the attack on Zhenjiang and the advance on Nanjing as part of the First Opium War. His dexterity with the Chinese language saw him enter and quickly progress through the diplomatic ranks to such a degree that he was instrumental in negotiating the Treaty of Tientsin (1858) at the conclusion of the Second Opium War. During this time he undertook a more systematic approach to translating the Chinese language into English, which was first promulgated in 1859. After forty years work in China, Wade returned to England permanently in 1883 and brought with him a vast, eclectic collection of Chinese material, including nearly 4,500 volumes of Chinese literature, which he donated to Cambridge University. It was largely on the strength of this that he was elected the first Professor of Chinese at Cambridge in 1888, a position he occupied until his death in 1895.

During Wade's later years in China, another scholar of equal erudition emerged from the ranks of the British diplomatic service; a man of very different stamp and one who would oppose much of what Wade had stood for in China. Herbert Allen Giles (1845–1935) became China's champion amongst the British. He developed an innate sympathy with and understanding of the Chinese race, and countered much of the prejudice and disdain that characterised the imperialist view of China, then prevalent amongst his colleagues in the diplomatic service and the business community they supported. Giles offered comparisons between China and the contemporary societies of Europe, comparisons that were generally unflattering to the latter. He vehemently opposed the work of the missionaries, arguing that the Chinese had no need of another religion and that

The confluence of the Xiao Jin and Fubian rivers near to Xiao Guan Zhai are marked by an unusual bridge constructed with steel chains (right) and a chorten perched on the hillside (above). Both indicate its strategic importance in past times.

The town of Xiaojin, Wilson's Monkong Ting, which he described as being 'most picturesque and strategically very strong'. The important position that this place occupied is still evident today.

evangelism came with too many strings attached. His strident position made him many enemies, not least within the consular service in which Thomas Wade was his senior official. Giles's forthright views did much to stifle his progression within the diplomatic ranks to the extent that he eventually 'resigned' on health grounds at the early age of 47 in 1893. Despite the many powerful people Giles had antagonised the strength of his intellect prevailed and in 1897 he was elected to the chair of Chinese at Cambridge, which had lain vacant since the death of Wade two years before. Giles had lost none of his feistiness and he used his influential position to castigate other sinologists whose work he considered inferior. He also besmirched the reputation of Wade, who he had always regarded as self-seeking and egotistical. He then set about improving Wade's transliteration work and it is the final irony that these two men, so estranged in life, should be forever linked in their system of Romanisation: the Wade–Giles system.

Our difficulties in the Xiao Jin Ho Valley resulted from the fact that Wilson had used the Wade–Giles system and the more rudimentary so-called Post Office system to transliterate the towns and villages he passed through. Unfortunately, the new communist government of China rejected both these systems, instead developing and finally adopting their own system – hanyu pinyin, literally meaning 'Chinese language spell and sound' – in 1958. This is now the official system and the one adopted internationally, save for a few locations, most notably Taiwan where the Wade–Giles system persists, more as a statement of its independence aspirations than as a practical preference. As a result of the widespread modern use of pinyin, all recent maps appear to have quite different geographic names from those used by Wilson; though in reality few have actually changed at all, their spelling is simply different. Consider the river Xiao Jin itself: to Wilson, using Wade–Giles, this was the Hsaochin, with the 'h' largely silent; today pinyin uses 'x' to represent the syllable 'sh', therefore Wilson's men would have used exactly the same pronunciation as our Chinese colleagues did. These changes also explain, to a large degree, the difficulties encountered by less well informed researchers looking for Wilson's itineraries. It is also interesting to note that there is no evidence that Wilson himself ever learnt the Chinese language at anything more than a superficial level, and was anything other than aware of some of the basic phrases that he needed to get by.

Wilson finally enjoyed some good weather on this part of his journey, as indicated by a photograph he took on 29

Wilson was much taken with the inhabitants of Xiaojin, commenting on their good manners and courteous behaviour. Today's inhabitants are equally friendly and welcoming.

June (see p. 100) looking forward in the direction of his travel, west up the river valley towards Danba. The sky is largely cloudless and shadows give evidence of a sunny day. As we drove down the valley we looked continually backwards to try and locate the position. Eventually the configuration of the hills above the valley seemed to assume an arrangement similar to Wilson's image and as we disembarked it was clear that we were very close to the location, but on the wrong side of the river. Fortunately, a recently built bridge allowed us to cross to the north side and line up what was a strikingly similar view, with a few key exceptions. Wilson's photograph has an upright poplar tree, heavily lopped, as its central feature and to the right habitations and carrying coolies resting along the rough riverside track can be seen. Not surprisingly the poplar had gone and the buildings had been demolished, though it was easy to see where they had been located. A local man approached us and provided lots of interesting information, the most telling of which was the fact that the route to Danba had been relocated to the other side of the river in 1980 when the tarmac road had been constructed. Therefore, once again we were literally standing where Wilson had stood and the nature of the routes he had traversed became all too apparent; then as now the rough, hard-baked and potholed track could barely be called a path.

Suitably buoyed we rapidly pressed on, passing the various villages that Wilson mentions until we reached a point where the Xiao Jin Ho is joined by another river from the north, the Fubien Ho (see p. 102). A little further up the valley was the village of Xiao Guan Zhai (Wilson's Hsao Kuan Chai), which was one of the places where the local tribes held up Qianlong's powerful Manchu army, as described in the last chapter. We were also told that the Red Army mustered at just this point during the Long March; clearly this section of country, though seemingly undistinguished, has had a great strategic significance over the centuries. At Xiao Guan Zhai Wilson had taken a striking portrait of a spreading walnut tree in a field of wheat with farm buildings in the background, but despite a diligent search we could find no evidence to locate this shot. More disappointing was the problem encountered a little further on at the fortified settlement in Dawei, with its characteristic solid, square tower. Wilson had taken an accomplished photograph of the village, successfully capturing the essence of the place. Though it was clear that we were at the same place we just could not locate where Wilson had pressed his shutter, even

'*Juglans regia* Linn. Tree 60 x 12 feet. Kuan-chai, near Mon-kong Ting, W. Szech'uan. 9,000 feet. June 27, 1908.' We hoped to be able to locate this striking tree and were greatly disappointed when despite a diligent search this proved unattainable.

260. Juglans regia Linn. Tree 60 x 12 feet. Kuan-chai, near Mon-kong Ting, W. Szech'uan. 9,000 feet. June 27, 1908.

applying our now well-established criterion of finding the most obvious and easy position. A great shame as Dawei still retained an air of defiant malevolence, its people regarding us with a palpable sense of proud independence. Wilson himself cautioned his readers:

> *the village of Ta-wei a considerable place for this region boasting a large lamasery. This place has an evil reputation, but no ill-will was displayed toward me. Many Lamas clad in claret-coloured serge crowded around and watched me as I photographed the village and displayed much interest in my camera, dog and gun. Nevertheless, the reputation of this village is well founded and I would advise travellers to avoid staying overnight here.*[3]

We took his advice, ending the day's journey at Rilong and the modern Siguniangshan Hotel. By the roadside the Xiao Jin Ho was a mere rivulet as it began its tortuous journey westwards to Danba and its rendezvous with the Da Jin Ho. I was saddened to finally leave the river behind; the journey up the valley had been every bit as interesting as we had hoped.

The Siguniangshan Hotel was named for the four mountain peaks that dominate the local area; the Four Girls Mountain, part of the larger Qionglai Shan complex. The largest peak of the Siguniang Shan includes Sichuan's second-highest mountain, which at 6,250 m is junior only to the mighty Gongga Shan. The whole area, over 1,375 km² in extent, was first opened to trekkers and climbers as recently as 1981 and, particularly on the northern side, remains pristine. In that first year a team of Japanese climbers conquered the high peak along its eastern ridge; every year since new attempts and new routes attract mountaineers to the area. In 1996 Siguniang Shan became a national nature reserve and some basic infrastructure was created; an entrance plaza and shuttle bus service to the main tourist site located in one of the high corries. We had sufficient time to take the bus into the park and were quickly swept up to above 4,000 m. Along the way the water meadows were filled with primulas, each meadow displaying a different hue as various species vied for attention. Ahead we saw low cloud and hints of snow-clad peaks and hanging glaciers. The bus terminated at a shanty town of stalls where local Gyarong people hawked their wares, mostly a pastiche of ethnic fakery. Tony and I had a few hours to fossick amongst the subalpine forest, composed of the last fragments of larch (*Larix potaninii*), and to scan the forest floor for any interesting plants. Sadly, most of the immediate area was badly eroded by the

The village of Dawei today. Frustratingly, it proved impossible to locate Wilson's position and take a matching image.

One of many striking water meadows within the Siguniang Shan National Park. The sheets of *Primula involucrata* subsp. *yargonensis* provide a watercolour wash to the landscape.

The exquisite beauty of *Primula involucrata* subsp. *yargonensis*.

passage of many visitors' feet. Despite this the air was invigorating and the views above, with enormous glistening snowfields, were breathtaking. Once away from the tourist area nobody was about and we enjoyed the solitude that makes these places so compelling. The whole scene was magical and evoked a sense of place more in keeping with the original Tibetan name of the mountain complex, Kula Shidak: Abode of the Mountain Protector.

The twentieth day of June dawned with clear blue skies as the sun climbed above the horizon. We left the hotel after a quick breakfast and continued heading east. Wilson endured overcast and humid weather with occasional rain, as he arrived at Rilong having scaled the Balang Shan Pass. His misfortune became immediately obvious as we climbed out of the village – the old part of which, known to Wilson, was now enveloped by modern hotels and shops – and into a layby at the roadside; looking back the whole panorama of the Four Girls Mountain was revealed, bathed in brilliant sunshine. Such an occurrence must happen very few times in the year and we felt privileged to witness it. Wilson is mute on this subject despite passing this exact spot and the only conclusion is that he saw nothing save for swirling cloud, indeed he makes no special reference to any mountains in the Qionglai Shan save for the Balang Shan itself.

The spectacular Siguniang Shan range or 'Four Girls Mountain', with the graceful *Larix potaninii* in the foreground.

From this point the drive to the pass was uneventful and we stopped a few hundred metres below to allow Tony and myself to climb the last section. As with all the high passes the meadows were lit by wonderful alpine flowers with the lampshade poppy once again a star performer. There was no sign of the inn where Wilson stopped for lunch on 24 June, not even a few courses of masonry. Such buildings were hastily erected and just as quickly abandoned. He named the place as Wan-jen-fen and other than describing it as a 'miserable hostel' makes no other comment. Sir Alexander Hosie is more forthcoming. He stayed overnight in the hostel in October 1904 and tells us the name of the hostel means 'the grave of 10,000 men', in remembrance of the Chinese soldiers who were killed near here fighting the local tribes.[4] These unfortunates were then buried in a mass grave. Hosie claims that a pavilion had been built containing a tablet with the names of the soldiers who fell. Wilson makes no such reference.

Wilson's party tackled the pass in miserable conditions:

Making an early start we toiled slowly over the dreaded Pan-lan shan, crossing the pass in a dense, driving, bitterly cold mist. The ascent is nowhere difficult and none of us suffered seriously from the effects of the rarefied atmosphere, in spite of the evil reputation this pass has from mountain-sickness. The ridge is narrow, razor-backed, the summit being composed of sandstone with marble embedded, piled up at an acute angle and devoid of vegetation. Snow, unmelted from the winter, lay in odd patches immediately below the pass and on all sides there was much fresh snow. The dense mists prevented any extended view, but what little we saw of the region was bare and desolate.[5]

The actual pass appeared quite changed, not surprisingly the construction of a road, where previously there had been the merest hint of a track, had made a great deal of difference. It seemed that the new road had been excavated from the old track to drop its level to a more secure base, taking a big 'V' out of the hillside. Somewhat foolishly, we took our image on the west side of the pass whilst Wilson's was taken on the approach from the east side. Nonetheless, a very individual group of rocks on the right hand side (left on Wilson's photograph) gave us a reasonable assurance that we were still travelling on Wilson's route.

We didn't dwell at the pass and quickly began to drop down the road on the east side. Wilson stayed overnight at another hostel in the place he named as Hsiang-yang-ping. He took an image of the building, jerry-built and squat, arranged as a rectangle around a central space. It sits astride a rude mountain track, lost in a swirling mist. He tells us the building was part temple and part inn and, in one of his most telling and amusing phrases, that it was 'kept by a priest to whose clothing and person water was evidently a stranger'.[6] He placed the building at 2,500 ft (762 m) below the pass, and as we wound our way down on foot we came upon the remains of a building at approximately this elevation. The situation didn't look very convincing in relation to Wilson's photograph, and without any kind of backdrop to go on it wasn't easy to match the image. However, it was also difficult to believe that, in this remote location, the remains could be anything other than the same building (pp. 114–15).

As we lost altitude the conditions rapidly improved and as the mist and clouds present at the summit of the pass began to disperse we emerged into warm sunshine. We had a high sense of expectation at this point; knowing that one of the potential highlights of the trip was just ahead of us. Wilson's description says it all:

The flora of the grassy ride leading up to the Pan-lan shan pass is strictly alpine in character and the wealth of herbs is truly amazing. Most of the more vigorous growing had yellow flowers and this colour predominated in consequence. Above 11,500 feet

Primula chionantha subsp. *sinopurpurea*, one of the striking alpine plants to be found below the Balang Shan Pass.

51. Pan-lan-shan, Summit, showing snow bare cliffs and slate shale. West of Kuan Hsien, W. Szech'uan. 14,650 feet. June 24, 1908.

'Pan-lan-shan, Summit, showing snow, bare cliffs and slate shale. West of Kuan Hsien, W. Szechuan. 14,650 feet. June 24, 1908.'

The Balang Shan Pass is still as bleak, desolate and devoid of vegetation as it was when Wilson passed by here.

49. Combined hostel and temple near summit of Pan-lan-shan. West of Kuan Hsien, W. Szech'uan. 13,650 feet. June 23, 1908.

'Combined hostel and temple
near summit of Pan-lan-shan.
West of Kuan Hsien, W. Szech'uan.
13,650 feet. June 23, 1908.'

The probable site of Wilson's
hostel at Hsiang-yang-ping.

altitude the gorgeous Meconopsis integrifolia *which has huge, globular, incurved, clear yellow flowers, cover miles of the mountain-side. Growing on plants from 2 to 2¹/₂ feet tall the myriads of flowers of this wonderful poppywort presented a magnificent spectacle. Nowhere else have I beheld this plant in such luxuriant profusion. The Sikkim cowslip (*Primula sikkimensis*), with deliciously fragrant pale yellow flowers, is rampant in moist places. Various kinds of* Senecio, Trollius, Caltha, Pedicularis *and* Corydalis *added to the overwhelming display of yellow flowers. On boulders covered with grass and in moderately dry loamy places,* Primula veitchii *was a pleasing sight with its bright rosy-pink flowers. All the moorland areas are covered so thickly with the Thibetan Lady-slipper orchid (*Cypripedium tibeticum*) that it was impossible to step without treading on the huge dark red flowers reared on stems only a few inches tall. Yet the most fascinating herb of all was, perhaps, the extraordinary* Primula vincaeflora, *with large solitary, violet flowers, in shape strikingly resembling those of the common Periwinkle (*Vinca major*), produced on stalks 5 to 6 inches tall. This most unprimrose-like Primula is very abundant in grassy places. The variety of herbs is indeed legion and the whole countryside was a feast of colour.* [7]

Cypripedium tibeticum

This sounded like a feast indeed, but would this be the sight that greeted us? Wilson witnessed this scene on 23 June 1908 and today was 20 June 2006; we were certainly at the right time but had this display persisted over the intervening decades or had some natural or man-made disaster destroyed it? Perhaps the construction of the road had caused massive disturbance?

We needn't have worried. We passed through the now familiar lampshade poppy, then one by one the other species began to appear, at first in small groups and drifts and then eventually by the tens of thousands. The floral carpet was every bit as striking as Wilson described it and the three outstanding plants were those that had captured his attention. *Omphalogramma* (formerly *Primula*) *vinciflora* was a stunning azure blue growing hugger-mugger with the Tibetan lady's slipper orchid, its huge pouch-like flowers seemingly at odds with its small stature. In damper places the stunning Sikkim cowslip was breathtaking. We revelled in the experience and had to be almost forcibly taken from the hillside after several hours.

Our journey east and down from the Balang Shan Pass continued on. Below 3,000 m the forest began to return. In the understorey *Rhododendron balangense*, a recently-introduced species endemic to the area, was just finishing flowering as was *Rhododendron galactinum*, another species with a limited distribution that had been introduced into cultivation by Wilson from these very plants. The forest vegetation was rich and varied, as Wilson noted, 'the variety and wealth of bloom was truly astonishing and I know of no other region in Western China richer in woody plants than that traversed during the

Our concerns that the meadows to the east of the Balang Shan, breathlessly described by Wilson, might have disappeared were quickly allayed and if anything were more spectacular than he indicated.

(from left) *Omphalogramma vinciflora, Trollius yunnanensis* and *Galearis wardii.*

Rhododendron galactinum raised from Wilson's original collection (W4254) on the east side of the Balang Shan, growing in the Valley Gardens, Windsor Great Park.

Despite a reputation as an eagle-eyed collector, Wilson appears to have missed *Rhododendron balangense* on his ascent to the pass. It not until 1983 that the Chinese botanist Fang Wenpei described this species.

Salix magnifica, a most distinctive and striking willow, is common along the valley bottom of the Pitao River on the road to the famous Wolong panda reserve.

day's march'.[8] Quite a statement, but one that was well founded, as even in this country so rich in native plants this area stands out as species rich. Indeed the mountains of south-west China have been designated a 'Biodiversity Hotspot' by the American wildlife organisation Conservation International. The concept of Biodiversity Hotspots was proposed by the British ecologist Norman Myers in the 1980s and recognises the fact that small, unique areas of the Earth have diverse clusters of both plants and animals. Thirty-four such hotspots have been identified and though they now only cover 2.3 per cent of the world's land surface they are home to over 50 per cent of the planet's plants and 42 per cent of land vertebrates. Conservation International tell us that these mountains are, 'arguably the most botanically rich temperate region in the world'.[9] This diversity also extends to birds, mammals, reptiles and amphibians; with the giant and red pandas and the golden monkey the flagship species. In all this diversity I was anxious to find one plant in particular – *Salix magnifica* – a most unusual willow. The leaves of this species have been likened to a magnolia, they are certainly very different from any other willow and most people, on encountering the plant for the first time, take some convincing that it is indeed a member of the genus *Salix*. Wilson tells us that the plant was common in these woods and sure enough as we descended the valley it began to appear in the mix of trees and shrubs. It can be found in discerning collections in western gardens though it is rarely encountered. Wilson himself introduced it from rooted cuttings in 1908, presumably from this same area.

47. Pan-lan-shan Valley. View in ravine, foot of, showing bridge, mixed vegetation and many conifers. West of Kuan Hsien, W. Szech'uan. 8,500 feet. June 22, 1908.

'Pan-lan-shan Valley. View in ravine, foot of, showing bridge, mixed vegetation and many conifers. West of Kuan Hsien, W. Szech'uan. 8,500 feet. June 22, 1908.'

The same view today. The
survival of individual trees is
one of the most significant
features of the landscape.
June 20, 2006.

We had two images that we hoped to match, but the chances seemed slight. Though very striking and conveying the nature of the valley with its precipitous cliffs, they had few unique features. Every turn of the road hinted at being the place where Wilson set up his tripod, yet nothing stood out. Mr Wang insisted he knew both places, although on hearing this Tony looked at me doubtfully. Once again we were shamed by our scepticism: as we drove over a bridge and along the valley, the car pulled to a halt and Mr Wang pointed back down the road. He was absolutely right. Wilson's caption is prosaic: 'Pan-lan-shan valley. View in ravine, foot of, showing bridge, mixed vegetation and many conifers. West of Kuan Hsien, W. Szech'uan. 22 June 1908.' As we glanced from the photograph (p. 122) to the view in front of us it became clear that many of the actual trees silhouetted along the sides of the ravine were still to be seen and some appeared to have grown very little over the period of 98 years. With a little adjustment we took an image that can be superimposed over Wilson's almost exactly. The only real change was the more substantial modern bridge and tarmac road. The river, the Pitao Ho, is joined by many other streams, which have carved out steep-sided lateral valleys, and at Yin Long Gou (Silver Dragon

Wilson and Walter Zappey in 1908. Though apparently enjoying a harmonious relationship in China, Wilson makes scant reference to Zappey nor does he particularly acknowledge Zappey's achievements in the field of zoology.

Valley) Mr Wang stopped the car once again and turned to us with a smile, 'your second photograph'. Remarkably, like the previous image, things had changed little, individual trees could be seen, the bed of the stream was the same and in taking the photograph we knew that the match was almost exact. This lateral valley was spanned by a bridge called the Erh-tao qiao and again we were standing in Wilson's footprints: 'following the torrent through a narrow ravine for 5 li we reached Erh-tao chiao, where the torrent connects with a very considerable stream which flows from the Pan-lan shan.'[10]

Just beyond this point our roads diverged. Wilson went slightly to the north over the Ni-tou Shan Pass whilst we more or less continued due east. The road skirted the nature reserve at Wolong, famous as the primary giant panda research station in Sichuan. All around us the forests maintained their diversity with occasional very large trees. Gradually the steep-sided valley began to moderate and the country opened up. Abruptly we emerged into the lower valley of the Min River where it meets the Chengdu Plain; the end of this stage of our journey. Not just the day's end, but also the end of the first part of the whole trip. Ahead of us was Dujiangyan, Wilson's Kuan Hsien, where he began the 1908 trip with a caravan consisting of 'eighteen carrying coolies and one head coolie, two chairs, two handy men, an escort of two soldiers, my Boy and self, making a party of thirty all told.'[11] Add these up, assuming that the 'chairs' each needed two men, and the total is actually 29. Other writers have concluded, rightly, that the thirtieth person must have been Walter Zappey, a naturalist from the Harvard Museum of Comparative Zoology, who accompanied Wilson on his third trip primarily to collect specimens of the avifauna.[12] That Wilson makes no reference to him at all in his formal account of this journey is most extraordinary.

In taking stock of the last 2–3 days Tony and I could gain some satisfaction from the images we had been able to match and the knowledge that we had tracked Wilson with a high degree of accuracy all the way from Kangding on what had been, in my view, his finest hour in China: a journey of great significance over demanding terrain and through botanically rich country.

'View in Pan-lan-shan valley. Cliffs, 500–1,000 feet sheer, with *Pinus wilsonii* Shaw. West of Kuan Hsien, W. Sichuan. 7,200 feet. June 21, 1908.'

46. View in Pan-lan-shan valley. Cliffs, 500-1,000 feet sheer, with Pinus Wilsonii Shaw. West of Kuan Hsien, W. Szech'uan. 7,200 feet. June 21, 1908.

Silver Dragon Valley. At this point we were within a day of Wilson's visit 98 years before. June 20, 2006.

CHAPTER FIVE

DIG THE BED DEEP, KEEP THE BANKS LOW

As we emerged into the Chengdu Plain the temperature and humidity rose appreciably, the skies became overcast and a fug clung to the Min River and the low hills which created its banks. The enclosed nature of this unique area of China, with high mountains to the west and north, the hilly Red Basin to the east and the Yangtze Valley to the south, accounts for its peculiar weather conditions. Snow and ice are virtually unknown and a cool winter gives way to an oppressive and cloudy summer with rainfall throughout the year. Wilson knew the plain and its key towns and cities well and my observations concurred with his: 'There are no extremes of climate in this region. In summer the temperature seldom reaches 100ºF in the shade; in winter it seldom falls below 35ºF. It is humid at all times and essentially cloudy, more especially in winter, when the sun is rarely seen, owing to banks of mist'.[1]

Dujiangyan is located at the northern edge of the plain at a point where the Min River emerges from the higher country before eventually terminating in the lofty Min Mountains, which border the neighbouring province of Gansu. The Min is a substantial river and when joined by the Dadu River its volume is swelled still further, though their combined waters still remain a mere tributary of the mighty Yangtze River which they join at Yibin. Wilson is very informative and knowledgeable about the river systems of Sichuan. The name of the province acknowledges the significance of its natural waterways, deriving from the phrase 'Four Rivers'; the four being the principal tributaries of the Yangtze in Sichuan, the Jinsha, Yalong, Min and Jialiang. Regarding the name 'Yangtze', Wilson comments that:

> *I have never met a Chinese to whom this name is intelligible. I have read that the name denotes 'Son of the Ocean' and is applied to the section between Wuhu and the sea. This may be so, I have no knowledge on the point. Many local names are given to stretches of this river, but from Sui Fu, in western Szechuan, to its mouth it is universally spoken of by Chinese as the Ta Kiang (Great River), occasionally it is rendered Chang Kiang (Long River) of simply Kiang, meaning* The River.[2]

A grand stone staircase leading the visitor to the Two Kings Temple above the Min River at Dujiangyan.

This is very much the situation today, for although the Chinese know that the river is called the Yangtze in the West, to everyone I spoke with on the subject in China it is the Changjiang: the Long River.

Ferry boats on the Yangtze River
near the important port of Wanxian.
To the Chinese this great water
course is the Changjiang.

The Chinese have always regarded their rivers as being of the utmost utilitarian importance; in past history these natural water courses provided a means to access the mountainous regions of the west at a time when roads were nonexistent. They fostered trade and commerce and today they provide the power to generate electricity. Hydroelectric schemes abound throughout Sichuan and the Min River is no exception. Our approach to Dujiangyan was marked by the degree of activity generated by the construction of a local hydroelectric dam. The whole area, not surprisingly, had been significantly affected by this large civil engineering scheme which had completely altered the river valley and the nearby streams which feed into the main river. Along one such tributary Wilson photographed the village of Hsuan-kou on the second day of his 1908 journey to Kangding. He captioned the photograph 'Hsuan-kou. A typical riverine village in W. Szech'uan'. Sadly the new dam has consigned this village to history as the rising water levels behind the dam have flooded its location. It now lies in a watery grave under many metres of the swirling, murky river Min. Wilson's photograph shows an attractive pagoda on the right bank of the river and Xiao Zhong informed us that this pagoda had been dismantled and relocated to higher ground further along the road prior to the flooding of the valley. This was such a rare example of conservation that we asked to be driven to the site. Apparently the pagoda was built in the tenth century, hence the efforts made to relocate it. Though the building had been rescued it looked forlorn and lonely in its new location – removed from the site of its construction and sitting alone on a grassy hillside without context or appreciation – but it does survive.

21.   Hsuan-kou. A typical riverine village in W. Szech'uan. 3,000 feet.
June 17, 1908.

'Hsuan-kou. A typical riverine village in W. Szech'uan. 3,000 feet. June 17, 1908.' The most important feature of this photo is the pagoda on the right hand side of the image.

The relocated and refurbished pagoda has survived the fate which befell its home village.

As we entered the outskirts of Dujiangyan the quality of the road improved considerably, and our final approach was along a broad, tree-lined avenue. The city is a World Heritage Site because of the ancient irrigation system it contains and, just to the south, the presence of a sacred, much visited Daoist mountain, Qingcheng Shan. Indeed, this mountain was a cradle of China's ancient religion. During the Eastern Han dynasty (AD 142) Zhang Daoling, one of the so-called Celestial Masters – the three highest divinities in Daoism – lived and taught on Mount Qingsheng. In the following dynasties temples were built on the mountain, generally in areas of great natural beauty, and Qingsheng Shan became one of the five key sites for Daoist pilgrims. In all, the site covers over 120 km and contains a wonderful array of manmade and natural features such as peaks, caves and waterfalls.[3] As a result Dujiangyan is a top tourist destination in Sichuan, particularly with residents of Chengdu just 50 km away, and the local authorities have invested large amounts of money in its infrastructure. Their efforts seem to have been successful as everything appeared clean and well ordered.

The Flying Crane Hotel on the outskirts of Dujiangyan. The surrounding trees are evergreen members of the Lauraceae, which Wilson called nanmu trees.

We arrived at our accommodation, the Flying Crane Hotel, just as nightfall closed in. Xiao Zhong informed us that the hotel had been built on the site of the temple buildings that Wilson had photographed on 16 June 1908. Most of the buildings had been demolished about 20 years before but the trees which set the scene in Wilson's photograph were all retained. Though the light was poor and we didn't want to prejudge his assertion we found it hard to reconcile Wilson's image with the situation presented to us. The hotel turned out to be very comfortable and well appointed and we enjoyed our short stay and the traditional Chinese hospitality that was extended to us. The grounds were delightfully landscaped and tall, evergreen trees were a feature of the gardens. The majority of the trees were various species of *Machilus*, a genus in the family Lauraceae. These trees were a favourite of Wilson's and were a feature of the cultivated land of the Chengdu Plain:

> *Around the houses bamboo, oak, 'Pride of India', soap trees (Gleditsia), cypress and nanmu are the commonest trees. Nanmu is a special feature around temples. Several species of the genus* Machilus *are called nanmu, all agreeing in being stately, tall, umbrageous evergreens. The wood they yield is highly valued and the trees are particularly handsome.*[4]

'Kuan Hsien, W. Szech'uan. Temple with bamboo and namu trees, (*Machillus nanmu* Hemsl.). 2,700 feet. June 16, 1908.'
Is this the same place as today's Flying Crane Hotel?

36. Kuan Hsien, W. Szech'uan.Temple with bamboo and namu trees (Machilus Nanmu Hemsl.). 2,700 feet. June 16, 1908.

The famous Li Bing and his son Li Er Lang, to who the prosperity of the Chengdu Basin is principally attributed. This monumental work of art is situated in the centre of Dujiangyan.

Nanmu translates as 'southern wood' in recognition of the fact that these evergreen trees are essentially subtropical and most abundant in the southern Chinese provinces. Their straight, tall trunks composed of dense, hard wood were used in the construction of temples, often forming the entrance pillars. They also made very effective shade trees and as such were used in the courtyards of the wealthier farmers to provide relief from the hot summer sun. Wilson's enthusiasm for the trees was not shared by his first employer, Harry James Veitch, probably because Veitch realised that their lack of frost hardiness limited their saleability in England. Nonetheless, Wilson did succeed in introducing several, some of which proved hardy and became established in cultivation. The celebrated tree of *Machilus ichangensis* at Wakehurst Place, Kew's sister garden in Sussex, is one such example, which Wilson stood beside with Gerald Loder (later Lord Wakehurst) on a visit to the garden in 1920.

The following morning over breakfast Xiao Zhong informed us that a local television company was interested in our enterprise and wanted to film us. They followed us for the rest of the morning as we went about our business, discreetly recording our activities and occasionally conducting interviews. The first job was to assess the hotel itself. Certainly the buildings were completely different from Wilson's image, excepting a boundary wall. The trees, however, did seem to match and though fast-growing eucalypts had been planted fairly recently several of the older, stately nanmu still stood out and were very similar in outline to those shown in Wilson's photograph. We were, however, still a little dubious and it was only the insistence of Xiao Zhong that carried the case, after all he had a lot more local knowledge than we did.

With our considerations of the Flying Crane Hotel completed we drove back into Dujiangyan to investigate its famous irrigation system and to match Wilson's image of this ancient engineering works. The curious and respectful Wilson devotes a considerable number of pages to explaining how the complex irrigation system worked and to eulogising the genius behind its construction: 'The plain owes its abundant fertility to a complete and marvellous system of irrigation, inaugurated some 2,100 years ago by a Chinese official named Li-ping and his son'.[5] Li Bing was a regional governor of the Shu State, one of the seven powers that made up China during the pre-imperial era known as the Warring States period. Shu had a nearby rival kingdom to its north and east which was populated by another aboriginal people, the Ba; these two kingdoms ultimately became the province of Sichuan. Both remain mysterious

萬頃波光歸稼穡
四山雲氣保妖記

二王廟

廟貌巍巍

The Two Kings temple at Dujiangyan is a fitting memorial to Li Bing and his son.

and enigmatic. Excavations of both civilisations have revealed their rich and unique cultures but have failed to cast very much light on their origins. The state capital of the Shu Kingdom, Sanxingdui, situated north of Chengdu, has been systematically investigated by archaeologists to reveal a settlement covering over 12 km² in extent, enclosed by a city wall that was over 2.5 km in length and up to 5 metres high. Wilson knew of the Shu and Ba kingdoms and made a special effort to travel through the area that once made up the latter during June and July 1910, a journey he found richly rewarding. These ancient kingdoms were destroyed by a still stronger rival, the Qin, who united all of China for the first time in 221 BC under its megalomaniac emperor Qinshihuangdi, for whom the famous Terracotta Army was created.

Whilst the Shu Kingdom was still an independent state, Li Bing and his son, Li Er Lang, began their great feat of engineering. Prior to this the Chengdu Plain relied on precipitation alone for its water and occasional droughts led to devastating famines. Yet the Min River with its superabundance of water followed a tantalising course just to the north and west of the plain. Wilson takes up the story of the development of the irrigation system:

A superb old specimen of
Lagerstroemia indica adorns
one of the many courtyards
at Li Bing's temple.

The principle on which the system is based is simple in conception, but very intricate in detail. An obstructing hill call Li-tiu shan was first cut through for the purpose of leading waters through and distributing them over the plain. The passage having been excavated, the waters of the Min River were divided by means of an inverted V-shaped dyke, a distance above the canal into two main steams, the 'South' and 'North' Rivers as they are called. The waters of the 'North' stream are carried through the Li-tiu shan cut and after passing through the city of Kuan Hsien are divided into three principal streams . . . The 'South' River, which occupies the original bed of the Min River, is divided into four principal streams almost immediately opposite the Li-tiu Hill . . . This system of anastomosing canals, ditches, artificial and natural streams forms a complex yet perfect network. The current in all is steady and swift, the bunding secure and floods unknown . . . These famous irrigation works are perhaps the only public works in all China that are kept in constant and thorough repair . . . The motto of Li-ping, 'Shen tao t'an ti tso yen' (Dig the bed deep, keep the banks low) has become an established law in these parts and is rigorously carried into effect. Amidst so much that is decaying and corrupt in China it is refreshing to find an old institution maintaining its standard of excellence and usefulness through century after century.[6]

Wilson's description is still an excellent introduction to the complex water works at Dujiangyan, and to all intents and purposes Li Bing's system is still intact today and the prosperity and productivity of the Chengdu Plain still dependent on it. Wilson's last few lines are interesting as they are very close to those written by Sir Alexander Hosie in his report to the British Parliament published some six years before: 'it is a pleasure to call attention to a public work which, so rare a thing in China, has not been neglected by the passing of the centuries'.[7] This is also not the only occasion that Wilson almost paraphrases Hosie in something that seems close to plagiarism: a description by Wilson of the preponderance of goitre amongst the inhabitants of the villages of Xiao Jin Valley is drawn almost verbatim from Hosie.

Not surprisingly Li Bing was all but deified by a grateful people and this worship has continued unbroken down the centuries. The hillside above the irrigation works is replete with an extensive range of temples and ceremonial buildings centred on the magnificent Erwang (Two Kings) temple which is dedicated to Li Bing and his son. Wilson was greatly impressed: 'It is by far the finest example I have seen in my travels, and is probably not excelled by any temple in China. It nestles midst a grove of fine trees, facing the river on the side of a hill with broad flights of steps leading from terrace to terrace'.[8] He goes on to detail the statues of Li Bing, his wife and son contained within the Erwang Temple, and to comment about the standards of maintenance and the many interesting trees that adorn the various courtyards, including a pair of magnificent crepe myrtle trees (*Lagerstroemia indica*) trained into fan shapes. Tony and I enjoyed wandering amongst the many impressive buildings and paying due respect to Li Bing. We could see no sign of the fan-trained trees but a superb open-grown specimen of crepe myrtle was a feature of one courtyard as were several old ginkgos. The landscaping and planting of the site was of the highest order and demonstrated a rare mastery of the combination of hard and soft elements. As we took our leave of the main temple complex Xiao Zhong pointed to a large wall-mounted sign, which was clearly of some antiquity, its Chinese characters boldly advising 'Shen wa tan. Di zuo yan': Dig the bed deep, keep the banks low.

Having enjoyed a tourists' view of Li Bing's works and the various temples built in his honour we returned to our real purpose. Wilson took an image of the temples nestling on the hillside above the inner stream of the Min River on June 16 1908 at the start of his

Tony and Mark below the sign proclaiming Li Bing's motto – Dig the bed deep, keep the banks low.

16. Erh-Wang Miao. Taouist temple dedicated to the son of Li-Ping, originator of the Kuan Hsien irrigation works. Kuan Hsien, W. Szech'uan. 2,800 feet. June 16, 1908.

'Erh-Wang Miao. Taouist temple dedicated to the son of Li-Ping, originator of the Kuan Hsien irrigation works. Kuan Hsien, W. Szech'uan. 2,800 feet. June 16, 1908.'

The Erwang temple nestles amongst a more mature hillside of trees than in Wilson's day and the hillsides beyond are clothed in much dense vegetation.

summer journey to Kangding. In order to tie-up this journey before moving north, we needed to try and locate the spot where Wilson had stood. This turned out to be quite a task. The whole complex is spread over a series of steep hillsides connected by a network of paths and though it was overcast, indeed showers fell intermittently, it was humid and warm making the going quite exhausting. Nonetheless, by a process of elimination we managed to locate Wilson's position. Our image matched his quite closely although the river banks had changed in the intervening years and the hillside was significantly more wooded than in 1908.

One final task remained before we left this intriguing area: to visit the famous and celebrated Anlan bridge which fords the Min River just at the point where it is divided by Li Bing's dyke. When Wilson began his trip in 1908 the bridge was having an annual overhaul and it was closed to traffic. His party had to go further upstream in order to cross to the opposite bank by another smaller bridge. He does, however, leave us with a description of how the bridge looked at the time:

> *This most remarkable structure is about 250 yards long, 9 feet wide, built entirely of bamboo cables resting on seven supports fixed equidistant in the bed of the stream, the central one being of stone . . . Not a single nail or piece of iron is used in the whole structure. Every year the cables supporting the floor of the bridge are replaced by new ones, they themselves replacing the 'rails'. This bridge is very picturesque in appearance and a most ingenious engineering feat.*[9]

The Anlan bridge over the Min River is still a striking feature despite a significant re-engineering of its structure in recent years.

Today the bridge rests on four concrete pillars and is a steel and timber structure, though still very picturesque in appearance.

One interesting interpretation aid that caught our eye on the way back to the vehicle was a reconstruction of the revetment system used by Li Bing to create the all important water dividing dike. A series of large timber tripods driven into the river bed were used to support a secondary timber frame which in turn held in place woven bamboo pads, stone ballast was infilled behind the pads. At the front edge a kind of primitive sausage-shaped gabion – rounded stones held in place by a lattice-work of bamboo – broke the force of the water. No doubt the whole thing needed frequent renewal but given that all the materials were plentiful, as was the labour to undertake the work, such a requirement was no drawback and illustrated the thought and ingenuity that created this most altruistic of ancient engineering works and attested to the humanity, far-sightedness and technical skill of the great Li Bing.

A modern interpretation of the revetment system used by Li Bing to divide the waters of the Min River in order to create his irrigation system.

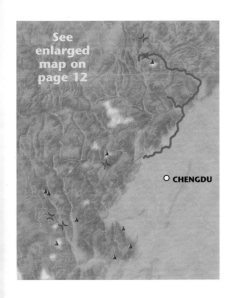

CHAPTER SIX

A MIDDLE WAY TO SONGPAN

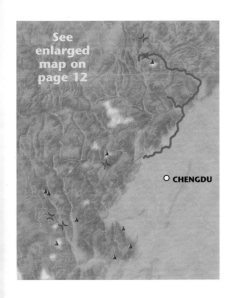

See enlarged map on page 12

○ CHENGDU

The regal lily was first discovered in the Min Valley by E. H. Wilson in 1903 and would later (in 1910) end his plant hunting career in China.

Wilson's fourth and final expedition to China, his second under the auspices of the Arnold Arboretum, was principally to gather seeds of conifers which had not hitherto featured extensively in his collections. This was partly because 1908, when he was in the Da Xue Shan and Da Pao Shan, had been a poor seed year. Charles Sargent specifically charged Wilson to make conifers his priority for this fourth visit to China.[1] Wilson himself also had some personal business. In 1903 he had discovered a lily in the Min Valley between Maoxian and Songpan and successfully introduced a small number of bulbs the following autumn. The bulbs flowered spectacularly in the summer of 1905 in the Veitch nursery.[2] A second, much larger, introduction to the Farquhar Bulb Company in Boston in 1908 had all rotted en route, much to Wilson's chagrin and largely due to his penny-pinching in not having them packed with sufficient care. Not surprisingly he was determined to put this right. Wilson had first considered that this lily was *Lilium myriophyllum* under which name it was originally distributed, thus causing considerable confusion as this name was also applied to *L. sulphureum*, a widespread species in Sichuan and Yunnan, and *L. sargentiae*, a plant found by Wilson in the adjacent Dadu Valley. However, he came to realise that it was, in fact, a new species restricted to a very small area of the Min Valley. He named this striking lily *Lilium regale*, the regal lily, and it was to figure large in Wilson's life.

By 1 June 1910, with Sargent's instructions regarding the gathering of conifer seeds in his mind and the ink on a new contract barely dry, he was back at Yichang in Hubei Province little more than a year after his return from his previous trip. We are not privy to Mrs Wilson's thoughts on this matter!

During June and July Wilson journeyed overland from Yichang to Chengdu. This was not his usual modus operandi, which was generally to undertake the river trip up the Yangtze and Min Rivers to Leshan to place him close to the mountainous areas of western Sichuan as quickly as possible. This change of approach can be explained by his eagerness to traverse the little known territory of eastern Sichuan, more for its historic and cultural interest than its botanical content as this heavily populated area had long since lost its original forest cover, and generally being below 2,500 m its weather was warm temperate: not a climate that would yield the hardy plants Wilson sought. Having successfully completed the journey it was time for Wilson to get down to business. He was very much at the top of his game, the complete traveller and collector. Having just forged this previously unknown route from Yichang he now decided that he would again take an off-road route to Songpan:

'*A few days after our arrival at Chengtu in 1910 I determined upon a journey to the border-town of Sungpan Ting, for the express purpose of securing seeds and herbarium specimens of certain new coniferous trees previously discovered by me in that region. During 1903 and again in 1904 I had visited this interesting town. On the first occasion I travelled by the ordinary main road, via Kuan Hsien and the Min Valley. The next year I followed the great north road across the Plain of Chengtu to Mien Chou, then travelled via Chungpa and Lungan Fu, by another recognised highway. On these journeys I gleaned tidings of a by-road leading from Shihch'uan Hsien across the mountains, finally connecting with both the above routes. This route promised to be interesting as well as novel. Only Roman Catholic missionaries had previously traversed it, so far as I could learn.*[3]

He provided us with another excellent photographic record of this journey, combining interesting views of the villages along the route, the various bridges that forded the rivers and streams, wayside shrines, occasional images of notable specimen trees and views of the beautiful Min Mountains, including their highest summit – Xuebaoding (Snow Treasure Peak). Tony and I had selected 23 diverse images that we hoped to be able match along the route. In many ways this journey was more important for its return trip than its outbound route, as it was at the point that Wilson neared Chengdu, within three days of the conclusion of the journey, that he had the near-fatal accident that was to bring an end to his time in China. Previous authors have vaguely referred to this as having occurred in the lower Min Valley; Wilson himself doesn't name the exact spot. Would it be possible to find, with some degree of accuracy, where this tragedy occurred?

This second part of our trip once again promised to be of the greatest interest and as we left Dujiangyan it was with a huge sense of anticipation. We had a very practical problem to face immediately. Wilson's middle route, then as now, was very much a byway; the two main northward routes to Songpan ran east and west of us respectively through Pingwu and Maoxian. Would it be possible to drive what in Wilson's day had been very much a footpath?

The regal lily growing in its natural habitat in the Min Valley, north of the village of Yanmen.

Wilson's portrait of his daughter Muriel in traditional Japanese attire. Opinion is divided on the merit of this aspect of his photography.

Things began auspiciously with a well-paved macadam road between Dujiangyan and An Xian. Our first image was an intriguing one. Wilson took many 'studies' – his attempt at using the camera not just as a faithful recorder of objects and scenes but as an aesthetic device to create artistic effects. Opinion varies as to how successful he was, many regard this aspect of his work as amateur and crass – his image of his daughter in traditional Japanese clothing gazing at herself in the mirror has been widely ridiculed. Others have been more measured in their opinion and regard his artistic attempts as showing a degree of sensitivity and depth.[4] At a place called Kung-ching-chiang, near Hien Chu Hsien (modern Mianzhu) on 10 August 1910, Wilson took a photograph of an attractive village temple, which he rather pretentiously titled 'A study in architecture' (see p. 242). It was such a singular building that, if it was still extant, we were confident it could be found. Along the way we stopped frequently to ask for directions, showing the image to reinforce our questions. We were directed to Xiang Fu Miao, a temple just outside Mianzhu. At first things looked promising but it quickly became clear that this was not the same building that Wilson photographed. Nevertheless we spent some time at the site, enjoying its ambience and the many attractive buildings. One temple building in particular was of great interest. Within its darkened interior, lit by a few slow burning candles, were 500 Buddha-like figures arranged around the four walls, each of which bore a different and very individual face. Some had kindly countenances; some wore bland expressions, yet others were ferocious. Each had very life-like eyes and in the reduced light they seemed to follow me around the room in a very unsettling manner. Despite my fascination I was happy to exit the building into the sunlit central courtyard. Xiao Zhong had been speaking to several of the monks and attendants and one old lady,

The temple building at Xiang Fu, though a false alarm, nonetheless exhibits some of the best qualities of Chinese architecture.

The main buddha with a selection of the 500 statues of deities that surrounded the temple interior.

0259. Bamboo suspension bridge 65 yards long. Near city of Shihchuan Hsien. A
2700 ft. Aug. 12, 1910.

'Bamboo suspension bridge
65 yards long. Near city of
Shihchuan Hsien. Alt. 2,700 ft.
Aug. 12, 1910.'

Today the situation looks quite different, though the hills and river banks identify the location as being the same as in Wilson's famous photograph.

who on looking at Wilson's image, had insisted that it was not a temple at all but rather a stage or outdoor theatre. Various nodding heads within the group indicated that this was the consensus view. Certainly the structure of the building with its floor elevated off the ground by legs and with screens and entrances would have made a good stage-set. With this interesting thought in mind we moved on.

We quickly passed through An Xian, a rather non-descript provincial town, and continued on a good road northwards. Beyond, An Xian's low hills began to appear, bare and denuded by centuries of human activity; according to Wilson coal mining was once an important activity. The whole scene was rather forlorn and uninteresting. Wilson crossed and re-crossed a river on his journey through the various villages: 'The river is broad and could easily be made navigable for boats during the high water season.'[5] Disconcertingly, all we could see was a dry stony river bed with the odd brackish pool, even though it was very definitely 'the high water season'. In discussing this with Xiao Zhong it became apparent that the valley had been robbed of its river to power another hydro-electric scheme. This development merely added another sad note to this melancholy area.

Though clearly on the route that Wilson had taken things became slightly confused. His sequence of towns – Lei-ku-ping, Che-shan and Shihch'uan Hsien – didn't seem to relate at all to the modern map or the situation on the ground, all the places we passed through were much bigger and full of modern structures; there was no sense of his passing here. To unravel the confusion required a good deal of discussion with Zhong. The town of Leigu was clearly Wilson's Lei-ku-ping, though unrecognisable from his description. The

The welcoming party of interested observers at the site of Wilson's bridge.

village of Che-shan and the important county town of Shihch'uan Hsien, by contrast, didn't seem to exist any longer. However, Zhong explained that due to its isolated position the latter had in fact been moved further down the valley in 1914 to the site of Che-shan and its name changed to Beichuan. As the new county town Beichuan had then expanded rapidly. What then of the site Shihch'uan Hsien where Wilson had stayed overnight on the night of 12 August 1910? The following day we discovered that this town was still very much in existence and to all intents and purposes little altered since Wilson's visit. It is now known as Yuli, in vague remembrance to a famous past resident – Dayu – who was a celebrated ancient engineer. With this sorted out, we settled into the rather dowdy Beichuan Hotel for the night.

The following morning we left Beichuan heading up the valley towards Yuli (now positively identified as Wilson's Shihch'uan Hsien). One of his better-known and widely published images shows a bamboo bridge spanning a river with a member of his party standing looking into the camera (p. 148). He describes this in detail in *A Naturalist in Western China*:

A few li below the city of Shihch'uan Hsien the river is spanned by a bamboo suspension bridge about 80 yards long, supported on cables made of split bamboo culms plaited together. These cables, eight in number, are nearly 1 foot in diameter and are fastened to stanchions fixed on either side of the river. Two similar cables on either side of the bridge are carried across at higher levels and have attachments of bamboo rope supporting those which form the base of the structure. A capstan arrangement is used for making the cables taut, and the lower ones are covered with stout wicker-work to form a footway. Like all such structures, this bridge is heavy, sags very much in the middle and is very unsteady to walk across. The life of these bridges is only a few years and strong winds often make them very unsafe.[6]

Wilson's image also shows a second similar bridge in the background spanning a side valley at right angles to the main bridge. He also took a second, side-on, photograph of the main bridge showing it against an adjacent hill topped by a small pagoda. Our hope was that we could find the bridge and match the images. Finding the location of the bridge was easy, it lay immediately next to the road we were travelling on. However, though the platform that once connected the bridge to the riverbank was still evident, everything else had long disappeared. A more modern concrete bridge had once again replaced the more picturesque bamboo bridge and it had been set further back up the valley. The secondary bridge had also been replaced, though in this case it remained in the same position as the old bridge.

The visit to Shihch'uan Hsien (modern Yuli) itself was very eventful for Wilson. Firstly, he discovered that he had been robbed:

Cash was needed, but on opening a box to obtain some silver for exchange we found that some one had stolen from it about 50 taels and 5 dollars. The load belonged to a coolie we had engaged at Taning Hsien, and retained because he had given unusual satisfaction! The previous day he had engaged a local coolie to carry his load on the grounds that he was feeling sick. He was last seen near Che-shan, still unable to carry his load. Evidently he was the culprit, but he was thoughtful enough to leave us about half the amount contained in the box. Since he had about three-quarters of a day's start I concluded it was best to quietly cut the loss, my first and last in China.[7]

This incident is interesting in revealing important elements of Wilson's character. As noted previously he was essentially a fatalist, particularly given his professional calling where danger lurked at every turn. Added to this he had to be hard-headed and practical in order

'Showing structure of Bamboo suspension bridge, laid on 8 cables, each a foot in diameter and suspended from two similar cables on either side. Floor of rough wicker work. Shih-chuan Hsien. Alt. 2700 ft. Aug. 12, 1910.'

0260. Showing structure of Bamboo suspension bridge, laid on 8 cables, each a foot in diameter and suspended from two similar cables on either side. Floor of rough wicker work. Shih-chuan Hsien. Alt. 2700 ft. Aug. 12, 1910.

The same view of the site in 2006 with the bridge platform still in evidence. In the background the second bridge has also been slightly re-located.

to successfully conduct his expeditions. He calculated that, at best, catching the miscreant would involve a significant expenditure of time and effort, if it were possible at all. Therefore, it was not worth the effort. This incident also reveals the essence of Wilson's dealings with the Chinese people, whom he regarded as essentially honest and fair-minded, and supports the consensus view that he was unusually successful in this regard; more so than his contemporary collectors many of whom had a disdainful attitude to the native people. Wilson chose to see the positive side of the incident and clearly never had any other such problems either before or after.

His tolerance of the Chinese people was tested again at Shihch'uan Hsien by the local magistrate, who had responsibility to ensure the safety of foreigners when in his district:

> *From Chengtu to this point we had travelled without escort, but with the difficulties of an unknown route before us I thought it best to secure such at the city. Sending my card to the Hsien's yamen in the ordinary way. I informed this official of my project and asked for the customary escort. Half an hour afterwards my card was returned with the information that there was trouble at Sungpan and no escort would be supplied! The refusal was as curt as it was insolent, but whether the Hsien was actually responsible I never found out. In my whole eleven years' travel in China this was the first and last experience of official discourtesy.*[8]

Once again Wilson took a resigned view of matters – much the best response in China, where even today official intransigence can be a maddening and frustrating experience.

Interestingly, the requirement for local magistrates to safeguard British citizens regardless of where they were in China was a consequence of the signing of another one-sided treaty between the two countries. The Chefoo Convention was negotiated and signed by Sir Thomas Wade (the self-same Wade of Wade and Giles fame – see Chapter 4) on behalf of the British Government in 1876. It followed the notorious 'Margery Affair', a diplomatic crisis caused by the murder of the British diplomat, Augustus Raymond Margery, and his staff in Tengyue in Yunnan in 1874. Margery was engaged in reconnoitring possible trade routes when he met his death. In an atmosphere of self-righteous indignation Wade forced the Qing government into a wide range of unconnected concessions, one of which was an extraterritorial right to protection for Her Majesty's subjects when travelling in the Chinese Empire. Wilson actually tells us that such protection was generally unnecessary or a decided burden, though on this occasion he clearly thought it prudent.

As if to try his patience still further, he had to endure a night at a very unpleasant hostel: 'We found accommodation in a large, curiously constructed inn remarkable for the strength of its stinks and the abundance of vermin and mosquitoes it sheltered'.[9] For any traveller in China the often objectionable nature of the accommodation could be very demanding. Each had their own worst experience. Sir Alexander Hosie wrote with feeling on the matter: 'nothing depresses the traveller in China at the end of the day more than the filthy accommodation provided by the ordinary Chinese inn. Quarters more than usually filthy are my fate tonight at Lung-ch'ih-chang, the first stage south-west of Omei Hsien.'[10] A. E. Pratt, a wealthy amateur naturalist who travelled through Sichuan in 1889 and 1890, had his worst experience at Lengji, home of the old ginkgo. 'I have been forced to take refuge in some curiously dirty places but this inn will remain fixed in my mind as containing the most varied collection of the most disagreeable things that I have ever met with at one time.'[11] For the record my own worst experience was in a forestry station near Taoyuan in the Micang Shan in October 1996 that had more than its fair share of stinging and biting creatures, and my miserable situation wasn't improved by the fact that I was suffering from an acute gastric problem at the time.

'Ornate memorial stone erected in memory of highly respected widow at Kai-ping-tsen. Shih-chuan Hsien. Alt. 3,200 ft. Aug. 13, 1910.'
This handsome memorial stone was a victim of the Cultural Revolution and pieces of it were purportedly incorporated into a nearby bridge.

0264. Ornate memorial stone erected in memory of highly respected widow at Kai-ping-tsen. Shih-chuan Hsien. Alt. 3200
ft. Aug. 13, 1910.

A tearful Mr Liang and his wife in their home with Wilson's photograph of his great grandmother's memorial stone.

We pressed on. The quality of the road began to deteriorate as we continued to drive north. The smooth macadam surface had been left behind and we struggled along on a pot-holed and muddy surface for several miles before arriving at the village of Kaiping where Wilson stayed on the night of 13/14 August 1910: 'a new empty house afforded us comfortable lodgings; the people were courteous and made our brief stay with them very pleasant. A remarkably fine headstone, recently erected over the tomb of a much-respected widow, was the chief thing of interest in the village'.[12] In the pouring rain we entered the village on foot. We had a copy of Wilson's photograph of the headstone and hoped to find it. Sadly, this was not possible, it had been destroyed by an act of vandalism during the Cultural Revolution in the 1960s but what emerged in the village provided a fascinating story. Many of the houses appeared to be quite old and it seemed possible to us that the village itself had altered little in the last hundred years.

Our appearance generated a good deal of excitement amongst the locals and Xiao Zhong engaged several of them in conversation. Very quickly a tragic personal history began to unfold from behind the image of the widow's headstone. Her family still lived in the village and we were taken to meet her great grandson, Liang Xue-Fu. As a young man he remembered the headstone and was delighted when we presented him with Wilson's image. He stood proudly with his wife for photographs before relating his family story. At the time of Wilson's visit his ancestors were rich and well-educated. They had come south to Sichuan from Shaanxi Province some generations before. His great grandfather, Liang Wei-Dong, had wished to join the Imperial Civil Service and throughout the dynastic period it was possible for men even in remote parts of the Empire to rise into the civil service ranks by passing a series of examinations. This system had been formalised into a three-tier structure during the Ming Dynasty. The first test was held in a local town and successful candidates became 'Superior Talents' and went on to a provincial level examination to become 'Promoted Scholars'. The top-level examination took place in Beijing and was called the 'jinshi'. This demanding test was based on knowledge of Confucianism, Chinese history and literary classics. Success in this examination would mean a lucrative government post either centrally in Beijing or in one of the provinces. This system was only abolished in 1905 just prior to the collapse of Qing Dynasty.[13]

Liang Wei-Dong became a 'Promoted Scholar' after gaining the best result in the middle-tier examination. He sat the jinshi examination in Beijing but failed it three times; according to the family tradition this was because he refused to pay the bribe that officials expected during the final corrupt years of the Qing administration. He returned to Kaiping a broken man harbouring a deep resentment that affected his whole personality and health, and he died within a short period of time. His wife became the matriarch of the family and a much-respected figure in the village because of her kindliness, as suggested by Wilson. Her children erected the headstone in her honour shortly before Wilson passed through the village. As Liang Xue-Fu related his story it was clear to see that his own circumstances were very much straitened, clearly he no longer possessed the wealth of his forebears. He did, however, retain a certain independence of spirit and pride and we left him with a sense of satisfaction, having provided him with a photographic link back to his long dead great-grandmother. Just as Wilson had 96 years before, we also found the inhabitants of Kaiping to be courteous and we were given a friendly farewell.

Despite the continued rain and poor condition of the road we made the next village, Xiao Ba, very quickly. Little had changed from Wilson's time, though the hillsides and village surrounds were much more heavily vegetated; indeed, though we managed to match Wilson's photograph of the village looking up the valley of the Bai Cao River, recently planted poplar trees obscured the view. Wilson's description of Xiao Ba remains pretty accurate today: 'this village, all things considered, is of considerable

Tony on the bridge at Kaiping.

The use of bamboo baskets still remains the principal means of taking locally grown vegetables to market, much as they did in Wilson's day.

0257. The market village of Hsao-pa-ti with slate roofed houses, Bamboo suspensi
bridge. Shih-chuan Hsien. Alt. 3700 ft. Aug. 14, 1910.

'The market village of Hsao-pa-ti
with slate roofed houses,
Bamboo suspension bridge.
Shih-chuan Hsien. Alt. 3,700 ft.
Aug. 14, 1910.'

Recent tree growth in the foreground has somewhat obscured the view but the match with Wilson's photograph is still easy to make. Perhaps the key change is the replacement of the mud shale houses with more substantial brick built structures.

Masons splitting the attractive green slate to be found throughout the Bai Cao valley.

Dressed slate ready for transport to the burgeoning building industry of Sichuan.

size (about one hundred houses), with many farmhouses scattered around. The mountains are less rugged and steep and are given over to the cultivation of maize. The houses are low, built of mud shales and roofed with slabs of slate.'[14] Such is the quality of the slate within this valley that a cottage industry has grown up supplying dressed slate for sale, with a great deal of it sent all the way to Chengdu. The coaster on my office desk is a piece of greenish Xiao Ba slate, a gift of one of the local masons.

We attempted to follow Wilson's route up and over into the next valley towards the village of Piankou, but at this point the road gave out and it was clear that from here on it was a foot-traffic only route. With insufficient time and equipment to follow we reluctantly turned around and headed back to Beichuan to take the road to Pingwu. This would rejoin us with Wilson's route at a place he called Shui-ching pu, which connected with the main Pingwu-Songpan highway. This left a gap of about 50–60 miles and about 5 days of Wilson's itinerary. It was a great pity as Wilson's account shows that in those last few days he saw many interesting things in the intervening country, including an area boasting a large number of giant trees of *Cercidiphyllum japonicum* var. *sinense*: 'Stumps of decaying giants abound, one of these, which I photographed, measured 55 feet in girth! [. .] These stumps are relics of the largest broadleaved trees I have seen anywhere in

0268. Northwestern Szechuan. Meliosma. Tree 60 x 18 ft. with huge head. Maize in foreground. Lungan Fu. Alt. 3600 ft. Aug. 19, 1910.

'Northwestern Szechuan. *Meliosma.* Tree 60 x 18 ft. with huge head. Maize in foreground. Lungan Fu. Alt. 3,600 ft. Aug 19, 1910.'

China'.[15] As his party began to climb into the foothills of the Min Mountains other trees and shrubs also began to appear. Perhaps one day it might be possible to return to Xiao Ba and walk the missing section: are the cercidiphyllums still to be found?

We drove quickly back to Beichuan and then turned north-east picking up a fast, well-surfaced road to the large city of Pingwu. This city was founded over 2,000 years ago and has had many names during this long period. Wilson knew it as Lungan Fu. Our stay was short, by the following morning we drove northwards again to pick up Wilson's route at Shuijing. Within a short distance we arrived at Yeh-tang, where a previous reconnaissance visit by Dr Yin had shown that a magnificent specimen of *Meliosma beaniana* that Wilson had photographed was still growing. It didn't prove difficult to find the tree as, once again, it was located right by the roadside. We were

Tony and Mark at the foot of Wilson's giant meliosma at Yeh-tang. This was perhaps the most tangible link with the great man.

The villagers of Yeh-tang appear justifiably proud of their remarkable tree.

Foliage of *Meliosma beaniana* on the tree at Yeh-tang, showing the characteristic brown felty buds.

Wilson's pressed specimen of the Yeh-tang tree in the herbarium at Kew. *Meliosma beaniana* WILS 4607.

certainly not prepared for the size of tree, it had grown significantly since Wilson's day and we measured it at 4.93 metres in girth. As we stood admiring the tree most of the villagers appeared, including Feng Zhen Quan, a local man who seemed to be the self-appointed guardian of the tree. He returned to his house and reappeared with a small notebook from which he began to tell us what seemed to be fanciful stories about the tree.

Apparently, some years before, a villager had tried to lop a branch from the tree for firewood but the tree had resisted and the person ended up almost chopping off their own hand. In 2005 some of the villagers had witnessed the emergence of a very large snake from a cavity high up in the canopy, a supposed spirit from the tree. This story was vouched for by several of the other bystanders. It seemed that the tree had a special, almost sacred aura for the local people, which might account for its survival. A small shrine at the foot of the tree seemed to affirm this status. With everyone gathered together Tony took the opportunity to give them more information about the tree, its botanical name, the link with Wilson and the importance of conserving it for future generations. We left Wilson's image with Mr Feng, with a feeling that the tree was likely to enjoy many more years of life. I almost imagined Wilson peering from behind the trunk of the tree with a satisfied smile on his face.

0269. Northwestern Szechuan. Hsao-Ho-Ying, an old military encampment now of r
importance. Mountains beyond 4000-6000 ft. above town. Lungan Fu. A lt. 560C
ft. Aug. 19, 1910.

'Northwestern Szechuan.
Hsao-Ho-Ying, an old military
encampment now of no
importance. Mountains
beyond 4,000 to 6,000 ft.
above town. Lungan Fu.
Alt. 5,600 ft. Aug 19, 1910.'

Wilson's description of the surrounding cliffs and peaks is well indicated in this duplicate image.

Back on the road we were able to follow Wilson's route directly, confirming his various observations about the surrounding country. The next photocall was the village of Xiaoho. Wilson took a striking image of this place with high peaks behind and tells us that it was an old garrison town, though even in his day its importance had long since disappeared. We were able to take an almost exact image, which seemingly had altered little, the chief difference being the destruction of a small monastery that had been active in Wilson's day. The village was a kind of gateway into a narrow valley with a spectacular aspect:

> *The scenery in this gorge, for magnificent, savage grandeur, would be hard to surpass. The cliffs, chiefly of limestone, are mostly sheer and 2000 to 3000 feet high [. . .] The mountain crests and ridges are covered with spruce and pine. Now and again glimpses of vicious looking, desolate peaks, towering above the tree-line, were obtainable.*[16]

These cliffs mark the beginning of the Min Mountains and are quite as dramatic as Wilson indicated. We left the valley by a series of 12 hairpin bends which provided a safe, if exhilarating drive over the ridgeline; Wilson exited by an exhausting scramble up the precipitous hillside.

Tony congratulating Mr Wang on his success at locating the last image.

Orchids abound in the protected World Biosphere Reserve of Huanglong. *Cypripedium flavum* is amongst the most showy.

0292. View of Hsueh-po-ting snows. Peak snow-clad 21000 ft. or more. Bed of stream in foreground encrusted with lime. East of Sungpan. Fr. Alt. 11500 ft Aug. 22, 1910.

'View of Hsueh-po-ting snows. Peak snow-clad 21,000 ft. or more. Bed of stream in foreground encrusted in lime. East of Sungpan. Fr. Alt. 11,500 ft. Aug. 22, 1910.'

Though not an exact match, the striking peak of Xuebaoding remains as stunning as ever. This is the closest we could get to matching the image without walking on the encrusted limestone.

'View of tarns of "many coloured waters" formed part natural and part artificial from lime encrustations. Also shows effect of lime on vegetation generally. East of Sungpan. Fr. Alt. 11,500 ft. Aug. 22, 1910.'

The same view in 2006. Sadly the sheer number of tourists robs this site of the solitude and seclusion Wilson experienced.

A local porter carrying dressed timber to the higher levels of Huanglong.

A strikingly similar image from 100 years ago.

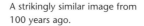

Soon afterwards we arrived at Huanglong: Yellow Dragon Valley. This amazing place is a Chinese Historic and Scenic Interest Area, World Heritage Site and World Biosphere Reserve. Such a series of accolades is well warranted, as this is a special site. Wilson begins his description in a rather understated way by describing it as 'a most interesting place'. The valley falls steeply from the snow clad heights of Xuebaoding, at 5,558 metres the highest point in the Min Mountains, over a relatively short distance creating a corrugated slope flanked by pristine mixed forests of larch, silver fir, spruce and birch. From the middle part of the year melting snow and summer rains bring water charged with lime down the valley, the lime precipitates out of solution to form travertine (tufa encrusted) pools and tarns each holding azure coloured water. With the giant mountain in the background a magical scene is created. To the vivid Chinese mind the valley appears as a long, sinuous dragon, its scaly skin formed by the ribbed and undulating slope. At over 3.6 kilometres in length this is a long dragon, but the illusion is easy to recognise and adds much to the mystery of the place. Not only is Huanglong a place of great natural beauty, it is also home to some of Sichuan's special animals including the giant panda and golden monkey. Within the forest Tony and I identified various rhododendrons, including *Rhododendron watsonii*, *R. rufum* and *R. oreodoxa*. On the ground, often very close to the alkaline-charged water, the Tibetan Lady's slipper orchid could be seen once again, this time accompanied by the related yellow slipper orchid, *Cypripedium flavum*.

We climbed up the valley on the solidly-constructed boardwalk, which keeps the many tens of thousands of tourists off the sensitive tarns. We passed the various features for which Huanglong is famed, each with a poetic Chinese name – Mirror-image Pond, Dragon Ridge Waterfall, Five-coloured Pond and Revolving Flower Pond. The walk passed a middle temple and terminated at an attractive upper temple and in looking back the whole of the valley was revealed. Across the other side of the valley a series of ridgelines, with razor sharp teeth emerging above the treeline, could be seen retreating into the distance. The Chinese liken this opposite hillside to a recumbent woman and call it the Sleeping Beauty. The whole area has a majesty and wonder that few other sites can match; it is indeed appropriate that Huanglong enjoys a twinning link with Yellowstone National Park in the United States, the world's first, and probably best known, national park. On the descent we matched two of Wilson's photographs, though it was difficult to keep the throng of visitors out of the camera shot.

0296. The hostel of San-chia-tsza. Alt. 12500 ft. one mile below summit of Hsueh shan pass. East of Sungpan. Aug. 23, 1910.

'The hostel of San-chia-tsza. Alt. 12,500 ft. one mile below summit of Hsueh-po-ting pass. East of Sungpan. Aug 23, 1910.

Tony standing within an ace of the position occupied by Wilson 96 years before.

Wilson enjoyed several hours here but didn't complete the whole climb: 'there was said to be another temple some few li higher up towards the snows, but I was too fatigued to visit it'. Nevertheless, despite his fatigue he had to push on towards Songpan. The night of 23 August 1910 was spent in a remote hostel that Wilson knew well:

The hostel at San-chia-tsze is maintained for the accommodation of travellers and a posse of soldiers is stationed here to keep down the banditti. The hostel is a roomy but miserable cabin, built of shales and roofed with shingles held down by stones. The floor is of mud and very uneven; there is no outlet for smoke save the doorway and no windows. At midday a candle was necessary to avoid falling over things when moving indoors. During different visits I have suffered many days and nights in this lonely spot, on one occasion being snowed in for three consecutive days.[17]

Unusually the image that Wilson took of this hostel includes himself standing in the foreground, hands in pockets staring pointedly towards the camera.

0300. The Hsueh-po-ting, 22000 ft. high, showing southwest face with snows and glaciers. From Hsueh-shan Pass. Northeast of Sungpan. Alt. 13300 ft.

'The Hsueh-po-ting, 2,200 ft. high, showing southwest face with snows and glaciers. From Hsueh-shan Pass. Northeast of Sungpan. Alt. 13,300 ft.'

An unchanged scene in 2006.

0298. View looking northeast from summit of Hsueh-shan Pass. Alt. 13300 ft, wit
ruined fort and town in foreground. Northeast of Sungpan. Aug. 23, 1910.

'View looking northeast from summit of Hsueh-shan Pass. Alt. 13,300 ft. With ruined fort and town in foreground. Northeast of Sungpan. Aug. 23, 1910.'

Though the fort has long since disappeared, the background crags and ridges identify the location as consistent with Wilson's image.

0299. View looking east southeast from summit of Hsueh-shan Pass. Eternal snows of Hsueh-po-ting in far distance, ruined fort with prayer flag in foreground. North-east of Sungpan. Fr. alt. 13300 ft. Aug. 23, 1910.

'View looking southeast from summit of Hsueh-shan Pass. Eternal snows of Hsueh-po-ting in far distance, ruined fort with prayer flag in foreground. North-east of Sungpan. Fr. Alt. 13,300 ft. Aug. 23, 1910.'

In 2006 prayer flags are still
a part of the landscape.

0297. "Moorland and crag" with tomb of man murdered by unknown bandits. East of Sungpan. Fr. alt. 13200 ft. Aug. 23, 1910.

'"Moorland and crag" with tomb of man murdered by unknown bandits. East of Sungpan. Fr. Alt. 13,200 ft. Aug. 23, 1910.'

The road over the pass has long since obliterated all remains of the poor coolie murdered here in 1910.

A parting view of the Xuebaoding
in the gathering evening light.

Various Chinese people, presumably including the innkeeper, are also in view. We knew from Dr Yin's previous visit to the site that the hostel had gone but both Yin and Mr Wang insisted the position of the building could be accurately located.

Having been proven wrong on several previous occasions we kept our doubts in check and our mouths shut and were right to do so. Nothing remained of the building at all, but a distinctive rock outcrop situated above the hostel on Wilson's photograph could be matched exactly, as could an adjacent section of scree. Tony took up Wilson's position and when the two images were later overlain he was standing just a metre or two to the right (p. 173). From the position previously occupied by the hostel a wonderful view of Xuebaoding could be seen and we were fortunate that the sunny conditions allowed an uninterrupted panorama. During the night of Wilson's stay he ventured outside:

> It was a perfect moonlight night on the occasion of my last sojourn at San-chia-tsze and late in the evening I beheld the 'Luck of Chengtu' with its crown of eternal snow lit by the radiant moonlight. The loneliness of the region and the intense stillness on all sides, and the wonderful peak with its snowy mantle made a most impressive scene.[18]

According to Wilson, when conditions were favourable, Xuebaoding was visible from Chengdu over 300 kilometres away. Such occasions were regarded as a lucky omen. All my Chinese friends insisted this could not be the case as Xuebaoding was much too far away and that the only snowy peak that could be seen from Chengdu was Xi Ling, a mountain in the Qionglai Shan range away to the south and west of us.

We followed the road up and over the pass, pausing to match several of Wilson's images (preceding pages) including further views of Xuebaoding and the surrounding ridges. In Wilson's time a ruined fort could be seen at the head of the pass and the grave of a recent murder victim: 'a poor coolie bound towards Lungan Fu to purchase rice was attacked here, robbed and killed. The bandits got clear away. The coolie's 'pai-tzu' (a framework for carrying loads on) and various appurtenances lay on top of the coffin and remain to tell the story of the crime'.[19] Not surprisingly nothing remained of either but their positions could be placed with accuracy and though the pass had gathered the usual shanty of jerry-built structures, providing for the tourist coaches, the essential peace and stillness that Wilson admired was still evident.

From the pass the road to Songpan could be seen tracing a thread-like line across the verdant mountainsides and we, like Wilson, 'sighted the city of Sungpan nestling in a narrow, smiling valley [. . .] with the infant Min, a clear limpid stream, winding its way through in a series of graceful curves.'[20]

CHAPTER SEVEN

NEMESIS IN THE MIN VALLEY

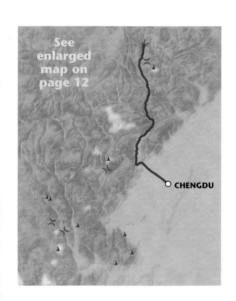

See enlarged map on page 12

CHENGDU

The almost ethereal beauty of the so-called red flag poppy – *Meconopsis punicea.*

As I accompanied Xiao Zhong to Songpan to sort out the hotel arrangements, Tony and Lao Yin drove in the opposite direction to the Gun Gan Lio Pass. Though they had no image to match Tony was anxious to visit the site where Wilson made his first collection of *Meconopsis punicea*, the red flag poppy. This had taken place on 31 August 1903.

Following his successful and speedy acquisition of the lampshade poppy Wilson had, in effect, time on his hands, such had been the ease with which he located the plant. Whilst studying herbarium material at Kew prior to his departure, Wilson had come across specimens of *Meconopsis punicea* collected by the Russian explorer Potanin during a journey through Gansu and northern Sichuan in 1885.[1] With time available Wilson determined to try to find this second species. To do so involved a stupendous journey from Kangding to Songpan, a distance of 885 kilometres (550 miles). Not only was the journey long and arduous but Wilson also suffered from repeated bouts of illness resulting in a weight loss of nearly two stone. The journey left an indelible mark on him and he provided the following brief description: 'I will not enter into any details of the journey. It was harder than anything I had before experienced. I was well-nigh exhausted in body, and almost in spirit, long ere the weary tramp was ended'.[2] Wilson was not a man to make such statements lightly; indeed it is rare that he speaks so candidly of his difficulties. Therefore, the demanding nature of the journey, which took 35 days, can be imagined.

On arrival in Songpan on 27 August 1903, his first visit to the town, Wilson wasted little time and within three days he was heading north to the Gun Gan Lio Pass (Wilson's Kung Lung Pass). The journey was easy and the poppy equally easy to find. Wilson reports that above 12,000 feet it was abundant, a situation that Tony confirmed when he returned to our hotel in Songpan that evening. The red flag poppy proved far less amenable to cultivation than its yellow counterpart. Wilson's original introduction was maintained in gardens for a few years and was supplemented by collections made by Joseph Rock in the 1920s but eventually all these died out and more recent reintroductions show few signs of being any more persistent. Even in favourable sites, with cool and humid conditions, this plant proves to be a short-lived beauty.

Songpan was Wilson's favourite town in Sichuan:

Did the Fates ordain that I should live in Western China I would ask for nothing better than to be domiciled in Sungpan. Though the altitude is considerable the climate is perfect, mild at all times, with, as a general rule, clear skies of Thibetan-blue. During the summer one can always sleep under a blanket, in winter a fire and extra clothing are all that is necessary. Excellent beef, mutton, milk and butter are always obtainable at very cheap rates. The wheaten flour makes a very fair bread and in season there is a variety of game. Good vegetables are produced, such as Irish potatoes, peas, cabbages, turnips and carrots and such fruits as peaches, pears, plums, apricots, apples and Wild Raspberries (Rubus xanthocarpus). Nowhere else in interior China can an Occidental fare better than at Sungpan Ting. With good riding and shooting, an interesting, bizarre people to study, to say nothing of the flora, this town possesses attractions in advance of all other towns in Western China.[3]

Meconopsis punicea is still abundant at the Gun Gan Lio Pass, the location where Wilson first discovered it in 1903.

The original settlement was most likely established by local people in antiquity but rose to prominence following Emperor Qianlong's military campaigns (as described in chapter three). It became a key garrison town and outward symbol of Chinese hegemony and power. It was enclosed by a substantial wall – over six metres (20 feet) thick and more high according to Wilson – and a general of the first rank resided there with a significant number of troops. It retained this importance during Wilson's time, indicating the general lawlessness of the surrounding area and the fact that it represented a last outpost of Chinese

The impressive North Gate at Songpan.

civilisation: 'Since the Chinese first established themselves here the town has undergone many vicissitudes. Time and again the Sifan have swept down upon it, captured it and massacred all who fell into their hands'.[4]

Tony and I first visited Songpan in September 2003 whilst engaged in a seed collecting expedition. At that time, its famous wall was largely derelict save for the impressive North Gate and the adjacent ramparts, which fully conveyed its past splendour. As a town of some size, within trekking country and not too distant from both the Huanglong Scenic Area and the larger and important World Heritage Site at Juizhaigou, Songpan was increasingly catering for tourists and several hotels were under construction. By 2006 things had developed significantly. The town wall had been rebuilt (and in some style), tourism was flourishing as evidenced by the number of hotels and there was a bustle about the place. Inside the wall the basic town layout, arranged around a cruciform road system, has probably not changed since the eighteenth century, but apart from one small section by the river, where old timber buildings could be seen, all the architecture was modern. Wilson tells us of a great fire that all but destroyed the town in 1901 and how it took the best part of ten years to make good the devastation. The Japanese also bombed the town during the Sino-Japanese War causing great destruction. The Min River sweeps right through the town, puncturing the east wall and necessitating the construction of several bridges, all of which are rather picturesque and elaborately built. Despite the passage of time and the great changes which have occurred in China, Songpan, like Kangding, has retained its essential character and ambience and it was easy to see why Wilson so admired it.

0310. View of city of Sungpan Ting from east, showing east gate and gap in wall through which river flows. Alt. 9200 ft. Sept. 25, 1910.

'View of city of Sungpan Ting from east, showing east gate and gap in wall through which river flows. Alt. 9,200 ft. Sept. 25, 1910.'

The river still follows its course through the centre of Songpan. The covered bridge can be seen in the far left of the photograph.

0312. City of Sungpan Ting from southeast showing military section of town. Al 9200 ft. Sept. 25, 1910.

'City of Sungpan Ting from southeast showing military section of town. Alt. 9,200 ft. Sept. 25, 1910.'

The town has long since
expanded beyond its
protective wall.

On the 1910 visit, Wilson entered Songpan on 23 August having taken 15 days to complete the journey from Dujiangyan. No doubt he was in good spirits, things were going well and he could look forward to several relaxing days before he began the return trip to Chengdu. He wasn't idle, however, and amongst other things he spent the afternoon of 25 August taking various views of Songpan from the slopes to the east of the town. We had investigated some of these images in 2003 as part of the growing realisation that it was possible to track Wilson's journeys; this time we had brought some additional images including one that looked north down the shallow valley of the Min River. We spent the rest of the day matching these images, tramping across the hillsides in the hot sun with the 'help' of a group of the local children, delightful urchins who ran around in front and behind us in a frenzy of excitement. The recent developments in the town and its surrounding area have made significant alterations to the scene, for instance there are now many more buildings outside of the wall than within. But the Min River still gently wends its way from the north and sweeps through the town with a flourish on its never changing journey south, the hillsides are still as denuded as in Wilson's day and the pastoral scene that entranced Wilson remains very much the overarching characteristic of this still remote corner of Sichuan. Interestingly, several images of Songpan are captioned as having been taken on 25 September 1910, clearly a simple error – August being the correct date – as by late September Wilson was ensconced in a missionary hospital in Chengdu in desperate straits.

The covered bridge seen by Wilson still remains the main access across the river within the city walls of Songpan.

Our Songpan 'helpers' on the
hillsides above the town.

The following morning Wilson left Songpan to start his return journey down the Min
Valley along a more established road, one he had travelled before. We have no published
account of this journey and have to rely on his journal and photographic record. It
began as a leisurely affair, the weather was pleasant, if a little hot, and Wilson spent the
time making observations about the country, its geology, vegetation and habitation.
Villages were strung along the road at regular intervals and he made steady progress,
averaging about 60 li (20 kilometres) a day. There was little to suggest that he was within
10 days of disaster.

We left Songpan just after dawn on 25 June intending to drive the length of the Min
River to Chengdu in a day. Despite being midsummer it was a raw morning, but as the sun
rose we could see it promised to be a fine day and with this most important section of road
ahead of us the anticipation levels were rising again. Our first image of the day was of a
crude bridge framed by two poplar trees. Wilson's caption, 'Min valley, South of Sungpan',
was far from helpful in locating the place and there was no reference to the trees in his
journal. However, within an hour we were there. The larger poplar was still extant and is
probably assignable to *Populus cathayana*, though Wilson records it as *Populus suaveolens* in
Plantae Wilsonianae, a species which is actually found much further to the east. W. J. Bean's
account of this latter taxon explains this confusion admirably.[5] As ever, a bridge was still
in position but was clearly of much more recent origin. It was a thrill to see the scene so
similar to how Wilson observed it and the poplar tree seemed to be in rude health having

0334. Cantilever bridge with two fine old Poplar (No. 4577) trees. Min valley. South of Sungpan. Alt. 8800 ft. Aug. 27, 1910.

'Cantilever bridge with two fine old poplar (No. 4577) trees. Min valley South of Sungpan. Alt. 8,800 ft. Aug. 27, 1910.'

Despite the very different light conditions, the poplar at the bridge at Xin Tan appears little altered.

gained significant height and girth during the twentieth century. The bridge allowed access to the village of Xin Tan – Wilson's Hsin Tung Kuan where he stayed on the night of 26/27 August. In taking an equivalent image we had two problems. One was the poor early morning light, which caused the poplar tree to be almost silhouetted against the bright blue sky high above the valley sides. It seems that Wilson also took his image in the morning, as it is dated 27 August, but it was probably not so early in the morning and, furthermore, the sky seems overcast in his photograph. Our second problem was the presence of a large lorry right in camera shot. This proved to be full of reconditioned refrigerators! We were told that the driver was asleep in the village, prompting Dr Yin to dash across the bridge to wake the poor man up. Within minutes he arrived looking decidedly sleepy and dishevelled, tucking his shirt into his trousers. With surprising good humour, he quickly fired up the vehicle's engine and moved it further down the valley allowing us to take additional photographs.

We pressed on. As the sun climbed into the sky the temperature began to rise sharply, the steep valley sides contained the gathering heat and created a close, uncomfortable atmosphere; not even a zephyr of wind relieved the oppressiveness. The road, previously running just above the level of the river, began to climb higher up the valley sides. The reason soon became clear as a wide lake opened up before us. From a vantage point just off the road we were able to take in the whole panorama. The lake was created when a violent earthquake, measuring 7.5 on the Richter scale, shook the valley on 25 August 1933. The attendant tremors caused huge rocks to break off the valley sides blocking the course of the river, which consequently rose rapidly. The ancient town of Diexi and all the other villages in the vicinity were destroyed by a combination of floodwater and rockfalls. There were at least 3,000 casualties in Diexi with only five people surviving according to

Diexi Lake: a reminder of the catastrophic earthquake which shook the Min Valley in 1933.

contemporary reports. It is said that damage and loss of life occurred in Chengdu and that the tremors were felt in Chongqing and Xian, respectively hundreds of kilometres away to the south and north-east. Forty-five days later the pent-up water burst through the dam-like blockage causing yet more casualties. Once the cataclysm was over Diexi Lake remained, connected via a stretch of the original riverbed to a second smaller lake. The remains of Diexi town lie under at least 80 metres of water. It seemed to me that there was a tangible sadness pervading the air as we surveyed the now tranquil scene; ironically, the area is today regarded as a beauty spot and is a favoured stopping point for coaches travelling between Chengdu and Juizhaigou. This whole area is tectonically active and earthquakes are a regular feature. Songpan itself suffered a significant earthquake – 7.3 on the Richter scale – as recently as 1976 and frequent minor tremors shake the valley on a regular basis.

The disruption to this area might account for the fact that we were unable to trace one of Wilson's more attractive images, an ornate covered bridge framed by trees of *Salix babylonica*. Once again there is no indication as to the name of the place. The date, 29 August, would seem to place him pretty much in the area of Diexi Lake and in all likelihood the bridge was a victim of the terrible earthquake; even Mr Wang looked blankly at the photograph. With the next image we were on firmer ground. Again, Wilson does not name the village, merely calling it a Sifan hamlet north of Hao-chou. We can be certain that it is the village of Shih Ta Kuan, today called Pai Shan Yin. Wilson's journal not only tells us that he stayed here on the night of 30 August, the same day that the village was photographed, but also gives us a good description of the place:

We stayed the night at Shih Ta Kuan (alt 6000') perhaps the most prettily situated village of the whole route. On the right bank the mountains rise sheer for fully 3000 feet. The village on the left bank is shaded by poplar, willow and walnut, and, though filthy as all the villages are, is most picturesque and the inn quiet and good.[6]

Wilson's photograph supports this assertion and today the village remains picturesque, though somewhat marred by the presence of satellite dishes and power cables that are now so ubiquitous throughout rural China. This is another place we had visited in 2003, the final confirmation for us that Wilson was more tangible than had hitherto been thought. On our first visit we spent some time in the village, mainly because we wanted to collect from the still existing walnut trees that Wilson mentions in his journal.

'Ornate covered bridge and *Salix babylonica* Linn. Min valley. South of Sungpan. Alt. 7,800 ft. Aug. 29, 1910.' This place proved impossible to locate and was probably destroyed during the earthquake in 1933.

0331. Ornate covered bridge and Salix babylonica Linn. Min valley. South of Sungpan. Alt. 7800 ft. Aug. 29, 1910.

0339. A Sifen hamlet, with grove of Walnut trees to right, Cornstacks to left
Min valley. North of Hao-chou. Alt. 7200 ft. Aug. 30, 1910.

'A sifen hamlet, with grove of
Walnut trees to right, Cornstacks
to left. Min valley. North of Hao-
chou. Alt. 7200 ft. Aug. 30, 1910.'

The modern village of Pai Shan Yin
retains the singular appearance
that so impressed Wilson.

A lady allowed us to knock some ripe fruits from her tree using a long bamboo pole that is kept for the purpose. She had two delightful children, a girl and younger boy. As we arrived for a second time the same boy appeared, bigger and more self-confident and happy to chat away to our Chinese team. His name was Xi Xiao Long (Little Dragon), a name that seemed somehow very apt.

We spent the next hour or so retaking the photographs that we had taken in October 2003, this time using a better digital camera and tripod to match things as accurately as possible. Being three months earlier, we also enjoyed the surprising diversity of flowering shrubs on the arid, stony hillside. Roses especially revel in the sunny conditions in all Sichuan's river valleys; many are highly local in distribution. Miss Willmott's rose (*Rosa willmottiae*),

Xi Xiao Long in 2006, just as friendly and inquisitive as in 2003.

Are the trees from which Tony collected walnuts in 2003 the same trees as in Wilson's 1910 image?

Our two young friends from 2003.

Rosa soulieana growing in the Minjiang Valley.

Flowers of the arid Minjiang Valley:
(*from left*) *Incarvillea arguta*, *Rosa wilmottiae*, *Cotinus coggygria* and *Koelreuteria paniculata*.

named for one of Wilson's key patrons, was resplendent on the hills around Pai Shan Yin, its charming dark pink flowers produced in great abundance. A close companion was *Rosa soulieana*, a more widespread species with white flowers and characteristic sea-green leaves. Also named for Ellen Willmott, the bright blue flowers of *Ceratostigma willmottiana*, were prominent in the understorey and grew in close association with *Incarvillea arguta*, a subshrubby species of this mainly herbaceous genus. The airy purple inflorescences of *Cotinus coggygria*, a common garden plant known as smoke bush, were also a significant feature of this rich habitat and further down the hillside a bright golden splash proved to be a big specimen of the golden rain tree, *Koelreuteria paniculata*, which obviously had some local significant as a small shrine had been secreted at its base.

Wilson continued his genteel pace blissfully unaware of the impending disaster that was about to befall him. He soon came upon one of his quarries. In the same way that Wilson had encountered it, the regal lily appeared for us with great suddenness within a few kilometres from Pai Shan Yin. Wilson records that '... the blossoms of this lily transform a desolate and lonely region into a veritable flower garden.'[7] With the

'The hamlet of Yen-Heng showing flat roofed houses. Sophora and Poplar trees. Valley of the Min River. Wei-chou. Alt. 4,500 ft. Sept. 2, 1910.'

0344. The hamlet of Yen-Heng showing flat roofed houses. Sophora and Poplar trees Valley of the Min River. Wei-chou. Alt. 4500 ft. Sept. 2, 1910.

The characteristic flat roofs in the village of Yanmen still remain.

temperature reading 36 °C in the narrowest sections of the gorge the lily could be seen sprouting from the rock faces with just a modicum of shade from patchy and now withered grasses. Though abundant, this lily has a very limited distribution, not just in being confined to the Min Valley but also in not occurring above or below a narrow altitudinal band of approximately 1,200–1,800 metres. Despite this it proved to be the easiest of plants to grow, adapting to a wide range of soil types and aspects, only really resenting deep shade and waterlogged conditions. Wilson recognised this combination of beauty and tolerance and declared '...in adding it to western gardens the discoverer would proudly rest his reputation with the Regal lily.'[8] Quite some accolade.

We had two images left to try to find in the Min Valley. The first one – a view of a typical village thoroughfare with curious locals gazing from within buildings at the camera with quizzical looks – proved elusive, again even Mr Wang couldn't help. The next image was easier. The village of Yen-Heng (today Yanmen) had characteristic flat-roofed buildings and lay in a curved arm of the river, narrow cultivated strips could be seen on the opposite hillside and a massive rock protruded from the left hand-side. Though the buildings were once again of modern construction they had preserved the flat roof form and other features of the landscape had been retained. Wilson's photograph was taken on the afternoon of 2 September 1910 and we realised that we were getting close to the point where catastrophe struck. As if to emphasise this Xiao Zhong told us that three days before our arrival a tourist bus had been caught in a rockfall, three passengers had lost their lives and 10 were seriously injured. We continued with a watchful eye on the almost sheer sides of the valley.

We approached the village of Sian Sou Qiao (Wilson's Soh Chiao) with a sense of trepidation. This was his last recorded overnight stay on 3 September 1910; after this his journal ends. His last few entries are interesting and strangely prophetic. He tells us that he left a man at Soh Chiao to mark lilies for later collection and to explore the forests over the ridges to the west. In all he expected 5,000 bulbs to be gathered. The last sentence of his journal, written in Wilson's own, at times almost unreadable, hand is all too clear: 'I am certainly getting tired of the wandering life and long for the end to come. I seem never to have done anything else than wander, wander through China'. Presumably this was written as he relaxed in Soh Chiao at the end of the day and yet it seems so uncannily predictive that I wonder whether it was written some time later, after his accident, to add a dramatic final flourish. We can never know.

The last entry in Wilson's Chinese field journal, on 3 September 1910, within which he appears to predict his own fate.

The following morning, 4 September 1910, began much as any other. Indeed, the company were in a buoyant mood, the journey's end being just a few days away. In the second volume of his book *Plant Hunting*, Wilson gives us a dramatic description of the events that unfolded:

Song was in our hearts, when I noticed my dog suddenly cease wagging his tail, cringe and rush forward and a small piece of rock hit the path and rebounded into the river some 300 feet below us. I shouted an order and the bearers put down the chair. The two front bearers ran forward and I essayed to follow suit. Just as I cleared the chair handles a large boulder crashed into the body of the chair and down to the river it was hurled. I ran, instinctively ducked as something whisked over my head and my sun hat blew off. Again I ran, a few yards more and I would be under the lea of some hard rocks. Then feeling as it a hot wire passed through my leg, I was bowled over, tried to jump up, found my right leg useless, so crawled forward to the shelter of the cliff, where the two scared chair-bearers were huddled. [9]

In *A Naturalist in Western China*, his description of the rockfall is almost laughably trivial: 'In 1910 when descending the Min Valley, I unfortunately got involved in a minor one and sustained a compound fracture of the right leg just above the ankle'. As understated as he may have been many years later, at the time he was in very serious trouble and began a flight for his life, utterly dependent on his team to conduct him with all speed to the comparative safety of the Friends Mission in Chengdu under the care of Dr Davidson and his wife.

As we passed through Sian Sou Qiao we guessed that he could have gone no more than five to six kilometres and once that distance had been covered Mr Wang pulled the vehicle to one side. He leant back to talk to us. We were at a place called Fu Tang Ba, a

notorious stretch of the route because of the frequency of rockfalls. Such was the danger that the road, which had originally run along the left hand or east side of the river, had been relocated via a bridge to the supposedly less hazardous right hand side. It seemed to us that this place was as likely as anywhere to have been where the great man had met his nemesis and the end of his plant collecting career in China, leaving him with a permanent limp; his self-styled 'Lily limp'. Appropriately on the opposite side, above where the incident probably occurred, thousands of regal lilies fluttered in the lightest of breezes almost as if Wilson had breathed a sigh of relief.

Our return to Chengdu was much less eventful than Wilson's and having negotiated the early evening traffic we were safely installed in the Philharmonic Hotel, near to Sichuan University. We had a few formalities to clear up before our return to England, not least to try

The regal lily on the hillsides above Fu Tang Ba, the probable site of Wilson's near fatal accident.

A roadside sign in the Min Valley indicating all too clearly the treacherous nature of the sheer-sided cliffs.

and locate some of the features that Wilson was familiar with during his regular stays in Chengdu. The city has a long and colourful history. First enclosed by a protective wall over 2,000 years ago, Chengdu has been at various times the capital for the Chinese Empire itself as well as an important regional centre. Due to the abundance of the Chengdu Plain it has always been a populous place, though not without its vicissitudes including war and revolt. The area is rich in natural resources and industry has been a significant element of the economy during the twentieth century. Wilson recognised the uniqueness of the city and admired the quality of the governance exercised by the local authorities, 'The city is clean and orderly, with an efficient police.'[10] He gives us an extensive and perspicacious account of Chengdu and its place in China as the curtain on Imperial rule was being drawn. He speaks with authority about the petty politics, the effects of Western influence and the likely future for the city and its people in the newly emerging republic. The account is of great interest both as a historical record and as an example of his insightfulness and range of interests; not for the first time Wilson shows us the breadth of his education and knowledge.

'House in Friend's Mission Compound, Chengtu, with Dr. and Mrs. Davidson. Alt. 1,700 ft. Dec. 20, 1911.'
It has not proved possible to locate this building. It is also interesting to note that the date on this photograph is wrong as it should read Dec. 20, 1910.

0348. House in Friend's Mission Compound, Chengtu, with Dr. and Mrs. Davidson. Alt. 1700 ft. Dec. 20, 1911.

The Jiu Yan Qiao spanning the Jinjiang River, a less than impressive replacement for the Marco Polo Bridge.

Chengdu was and is a dynamic place, a powerhouse of the western provinces. In the short time since my first visit in 1996 the city has changed out of all recognition. Modern roads, buildings and shops characterise the environment and have been built at a breakneck speed. A new, modern airport now serves this regional hub. Restaurants abound and today the residents of Chengdu are more likely to drive a car than ride a bicycle. With a population of 11.03 million people in its wider urban area Chengdu ranks as the fifth largest city in China behind Shanghai, Beijing, Tianjin and Chongqing. Not surprisingly there is little that Wilson would have recognised but we were interested to see if any of the key elements relevant to Wilson survived. These were Dr Davidson's Mission House, the Marco Polo Bridge, the city wall and Qing Yang Gong temple, which Wilson photographed on 18 August 1908.

Xiao Zhong, who was born in the city and has lived all his life in Chengdu, could provide no information about the Mission House, where Wilson was operated on and convalesced after his accident, and he suspected that it had long since been demolished, or, more likely, converted to other uses. He also knew that the so-called Marco Polo Bridge had been destroyed sometime in the twentieth century, an event that was a clear disappointment to him. However, he took us to the site of the bridge, more accurately known as Jiu Yan Qiao – the nine arch bridge. The new bridge retains the name despite the fact that it only has three arches. It spans the Jinjiang River and carries a heavy burden of Chengdu traffic in and out of the city every day. The magnificent city wall has fared little better. Wilson tells us: 'This wall is 66 feet broad at base, 35 feet high, and 40 feet broad at top, along which runs a crenulated balustrade. It is faced with hard brick (the walls of all other cities on the plain are sandstone) and is kept in thorough repair'.[11] It was originally over 12 kilometres in circumference. As a child Xiao Zhong remembered it as still being quite extensive, though ruinous. Today one small section exists, restored and well presented but still just a mere fragment.

The last remaining section of the once magnificent circuit wall of Chengdu.

The old street of Guanxianzi, a hint of the Chengdu that Wilson knew.

The Daoist temple has been maintained in pristine condition since Wilson's day and provided a fitting last match at the conclusion of our journey.

From the wall, we were taken to nearby Guanxianzi, an old street that gave some hint of the city that Wilson had known: 'The streets are always crowded with pedestrians, chairs and wheel-barrows. Different trades occupy their own particular quarter. Certain streets are devoted to carpentry in all its branches, boot-shops, shops devoted to horn ware, skins and furs, embroideries, second-hand clothes shops, silk goods, foreign goods and so forth'.[12] Guanxianzi presented a rather forlorn appearance, certainly it looked ancient but an air of neglect and decay pervaded the scene despite the fact that we were told it was being restored. It did, however, convey a sense of a time gone by and in my mind's eye I could picture Wilson sauntering down the street his eagle eye missing nothing, occasionally peering in a shop window to examine some exotic trinket.

Our final visit took us to the temple at Qing Yang Gong (Green Goat Temple). The photograph taken by Wilson was not one that we had brought along but one that Dr Yin had acquired. He was anxious to show us this famous Daoist temple and after a frantic drive across the city we arrived at a busy but little altered edifice. One last time we set up the tripod and matched the image. Fittingly we got a pretty close match and as Tony and I stood admiring the ancient building we considered what we had accomplished on our trip to find Mr Wilson. Though we hadn't achieved everything we had set out to do – important images had escaped us – we had certainly been able to cast a new light on his travels and scotch the myth that his journeys were untraceable, but most of all we had followed where Wilson had trod and relived some of the experiences of this most famous of plant hunters.

CHAPTER EIGHT

SAVING WILSON'S PLANTS

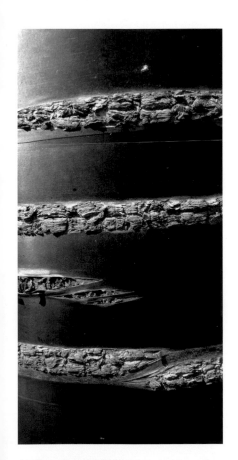

Another ornamental tree introduced by Wilson with spectacular bark is the Tibetan cherry, *Prunus serrula*.

Opposite: The paperbark maple, *Acer griseum* growing at Kew, introduced by Wilson in 1901. This is one of the best ornamental trees for showy bark.

Wilson's achievement in introducing so many new plants is well documented. W. J. Bean tells us:

> *Wilson's services to horticulture and his contribution to botany have not probably been equalled by those of any other collector. To give some idea of the magnitude of his labours, it may be mentioned that he introduced some 1,200 species of trees and shrubs, amongst which have been found 400 new species and 4 new genera.*[1]

In *The Hillier Manual of Trees and Shrubs* we learn that, 'he is said to have introduced over a thousand plants new to western cultivation, most of them woody and many still considered to be garden standards'.[2] Alfred Rehder, his co-worker at the Arnold Arboretum, provides us with the most authoritative account of Wilson's introductions. Rehder wrote most of the text for *Plantae Wilsonianae*, the botanical account of Wilson's Chinese introductions, and so he spoke from first-hand knowledge. Rehder provides us with the definitive obituary of Wilson in the *Journal of the Arnold Arboretum*, itemising in detail his collections of herbarium specimens, seeds, plants, photographs and publications. His final sentence is, perhaps, the ultimate accolade for Wilson's horticultural achievements: 'His name will live through generations to come in the new plants he discovered many of them commemorating his name and in the plants he brought from foreign lands to embellish our gardens'.[3]

Of equal importance to the actual number of plants Wilson introduced was their ornamental quality. Not only is China a treasure house of plant diversity with an estimated 30,000 native species, it has a high degree of endemism – plants found nowhere else. From within this exclusive group we find another remarkable phenomenon; in horticultural terms many Chinese plants are regarded as the most ornamental representatives of their particular genera. By way of example; whereas Europe has the field maple (*Acer campestre*), an unassuming and modest species, China provides the paperbark maple (*Acer griseum*), an exquisitely exotic tree. In North America we find the Jack in a pulpit (*Arisaema triphyllum*), charming but plain, whereas China gives us the more flamboyant *Arisaema candidissimum*. The British native rowan (*Sorbus aucuparia*) is a tough, adaptable small tree with heads of white flowers and masses of red fruits, much anticipated by the birds; its North America cousin, *Sorbus americana*, is a plant with a similar lack of pretentiousness. In the mountains of western China, by contrast, a legion of pink- and white-fruited rowans can be found, each striking in its beauty and ornamental merit.

This photograph of the fruits of *Actinidia chinensis* was taken by Wilson in China. He was an early sponsor of this climbing vine believing, correctly, that it would one day be an important food source.

Some, such as *Sorbus setschwanensis*, are delicate with filigree leaves and a tracery of branches, others, such as *Sorbus pseudohupehensis* are stronger-growing and give us stunning autumn colour with salmon and apricot tints and masses of pinkish fruits. The sheer diversity of this cluster of *Sorbus* species is staggering. This comparative list could fill a small volume in its own right, for, by whatever quirk of nature or act of God, China was endowed with not just an extensive natural flora but also one with garden worthy plants in abundance. Wilson was not slow to realise this.

Though Wilson had a botanical leaning, he was at heart a horticulturist; he knew a good garden plant when he saw one. This fact coloured his appreciation of plants and the approach he took to his expeditionary work. It is interesting that the plants he was most celebrated for in his lifetime were those with the most ornamental value: the regal lily and the dove tree. With the benefit of hindsight, however, there is a strong argument that his most important introduction was, in fact, the Kiwi fruit (*Actinidia chinensis*). This temperate Chinese liana is now regarded as a key health food providing important elements such as potassium, copper and magnesium. Rich in vitamin E and C, the latter an important antioxidant, and dietary fibre, today it is one of the principal agricultural products of New Zealand with an annual production value of over £250 million.

In considering Wilson's achievements, as represented by his plant introductions, we must not forget that he worked for two very different organisations during his time in China. His employment by the Veitch nursery was on a purely commercial basis. The hard-headed Harry James Veitch and his nephew James Herbert Veitch had one objective in mind, to add novel and commercially valuable plants to their catalogues. They were not interested in anything that didn't have a saleable value. Therefore, Wilson was under strict contractual instruction not to waste time or money on anything other than gathering ornamental plants. Sadly for Wilson the full credit for his collections for Veitch was never realised, not just because in the eyes of the autocratic Harry James Veitch and in the rigid social hierarchy of the time Wilson was a mere employee – 'our collector in China' – but because the whole Veitch nursery business was wound up in 1914. Many of Wilson's trees, in particular, were barely known at this point and were sold by the job lot in the great sale that was coordinated by Messrs Protheroe and Morris.[4] Rather than emerging to a horticultural fanfare his hard won plant treasures were announced by the auctioneer's gavel.

For the Arnold Arboretum Wilson had a freer hand in China but one that was directed more towards botany than horticulture. He was also hampered at the Arnold by the climate of Boston. The temperate forests of Hubei and Sichuan contained many plants that were ill-adapted to the New England winter and several of the most important groups were never planted or failed to establish in the Arnold Arboretum. Wilson's magnolia seedlings, for example, were sent to Léon Chenault in Orleans, France in 1913 not just to use the nurseryman's propagation skills but also to enjoy the more congenial conditions.[5] Nonetheless, Wilson had one distinct advantage at the Arnold: following his return from China he was taken on to the Arboretum staff. Initially this was to work through his Chinese collections with a view to writing and publishing a botanical treatise, what eventually became the three-volume *Plantae Wilsonianae*. Increasingly, however, he began to play a role within the living collections of the Arboretum itself personally setting out many of the plants that he had introduced.[6] This was an opportunity denied to any of his contemporary collectors and ensured he could continue to keep a direct interest in his hard-won plants, to watch them develop and learn their idiosyncrasies. This increased his knowledge and gave him the platform to write and lecture and advise others, spreading his renown still further. In his youth the celebrated dendrologist Sir Harold Hillier met Wilson who, not surprisingly, left a lasting impression on the young man:

> *Unlike so many academic botanists, Wilson was a really interesting companion with whom to tour a garden. He had a remarkable knowledge of plants, both woody and herbaceous, and he was never at a loss for the correct name. He was the sort of man I call a field botanist, and I always regarded Ernest Wilson and W J Bean as the two greatest professional plantsmen I have ever had the pleasure of meeting.[7]*

Wilson was good company in the garden. His colleague Alfred Rehder, by contrast, was less open. In the same article Hillier tells us that he would, 'seldom hazard a name when making a tour of the garden but would take a specimen and a fortnight later inform you of its identity'.

Wilson maintained an interest in his own plants to the end of his life. His final written work was a still unpublished manuscript detailing, on a genus by genus basis, the stories of his plant introductions and, more importantly, their performance in cultivation. This required a voluminous correspondence with all his horticultural contacts, in most cases the owners or managers of the most important gardens of the day on either side of the Atlantic. Though it was nearly 30 years since his first introductions had been made a great many of his collections had barely got going. All the stock purchased in the Veitch winding-up sale by Gerald Loder at Wakehurst Place, for example, had been planted for little more than 10 years. With a gap of a hundred years we can now take a more measured view of his Chinese introductions and their current status.

Several authors have written about Wilson's plants in general terms but few have analysed their impact.[8] If we consider Wilson's real mètier – woody plants – it seems to me that they fall into three loose categories. The first group, as suggested by *The Hillier Manual of Trees and Shrubs*, have become standard constituents of the garden flora or of civic landscaping schemes in the temperate parts of the world, though in many cases few people know them as Wilson plants nor recognise their origin. *Abelia schumannii, Clematis montana* var. *rubens, Cotoneaster dammeri, Kolkwitzia amabilis, Lonicera nitida* and *L. pileata, Pyracantha atalantioides* and *Viburnum davidii* are just a few examples. A second group contains other introductions which provided the genetic resources that allowed hybridists and nurserymen to cross and select whole new races of garden plants in genera such as *Deutzia, Magnolia, Rosa* and *Rhododendron*.

The final group have always remained plants for the more discerning horticulturist and tend to be found in botanic gardens, National Trust properties and the gardens of the cognoscenti where their virtues were fully appreciated. They contain some of China's most arresting beauties. Their lack of impact or widespread recognition is often hard to explain. Few present intractable propagation problems, nor are many capricious in their temperament. It would seem that their relative anonymity can only be explained by the limited appetite or knowledge of the average gardener, garden designer or landscape architect. Such a situation is a great pity. Perhaps the reader will therefore forgive Tony and myself a little self-indulgence if we each present our top ten Wilson woody plants.

The beauty bush, *Kolkwitzia amabilis*, a widely grown shrub, particularly in the United States, introduced by Wilson on his first expedition to China. Selections made in cultivation have produced superior cultivars such as 'Pink Cloud' and 'Rosea'.
Photograph © Mark Flanagan

MARK FLANAGAN'S SELECTION

In my own case these are drawn from an ever-changing selection which currently begins with *Acer henryi*, a small tree of great charm and distinction. I was fortunate enough to encounter, collect and successfully raise plants from a tree in the Micang Shan, north-east Sichuan, when travelling with Tony, Bill McNamara, the Director of Quarryhill Botanical Garden in California, and Charles Howick in 1996. It is a maple with trifoliate leaves which emerge with a reddish flush. It a rare species in cultivation and is difficult to find in the trade. This is a great pity as it forms a beautiful arching specimen, with leaves that turn a vibrant orangey-red on falling in the autumn.

Betula albosinensis is an inevitable choice, inevitable as it is one of the most stunning temperate trees not just for its bark, a subtle combination of pink, orange, red and cream tones washed, in young specimens, with a ghostly bloom, but also because of its clean, well-fashioned foliage of a unique sea-green colour and singular upright form. Wilson collected the tree many times but his best introduction is W4106 which was collected by his men in mountains above the Min River in October 1910. I have seen this tree throughout northern Sichuan where it is a striking component of the rich, mixed forests. The variety *septentrionalis* is supposedly a distinct entity and was first collected by Wilson in the Da Pao Shan. Having seen great variation in the wild I would contend that this is

Another of Wilson's show stoppers is *Betula albosinensis*, often called the Chinese red-barked birch. This tree never fails to delight when it is well-grown and multi-stemmed specimens can be particularly effective.

China has a very large number of exquisite maples, the *Flora of China* records 96 species as being native. Many well informed horticulturists regard *Acer henryi* as one of the best, though sadly it still remains a plant for the connoisseur.

part of the variable gene pool of this tree, a view shared by Alfred Rehder who had grave doubts about the validity of this variety. In horticultural terms, however, trees grown as *septentrionalis* are distinct with a much darker bark that emphasises the surface bloom on young trees. William Purdom's collection from Gansu is the most garden worthy and it should be sought out.

Catalpa ovata is the gem of a genus which is best known for the North American Indian bean tree, *C. bignonioides.* The latter is a rather coarse-leaved tree of ungainly habit whose saving grace is its panicles of showy, though fleeting, flowers in mid-summer. The yellow catalpa, by contrast, forms a tall growing tree with large but refined leaves and panicles of creamy-yellow flowers which have an elusive scent, faintly reminiscent of strawberry smoothies. Its long and slender 'beans' are held through the winter on leafless branches. Interestingly, Wilson was not the first person to introduce this tree, that honour falls to Philippe von Siebold who collected seed from cultivated trees in Japan in the 1840s. Wilson, however, gathered it from its native range in the woodlands of western Hubei where, he tells us, it was locally common.

The Chinese hazel, *Corylus chinensis,* is very different from our native hedgerow shrub and is a tree of the first rank reaching 30 metres in the forests of eastern Sichuan and western Hubei, from where Wilson introduced it in 1900. In time it produces distinctive flaky, almost cream-coloured, bark. Once again we found this tree in the Micang Shan holding its own within the climax woodland.

Catalpa ovata is the most striking of the three or four *Catalpa* species to be found in China. Large trees are rare in cultivation but they form spreading specimens up to 15 metres that lack the heavy, sombre feel of the better known North America trees.

A hazel up to 30 metres tall on a straight trunk with flaky bark sounds like a fanciful idea but this accurately describes the Chinese hazel *Corylus chinensis*, another of Wilson's remarkable finds. As a bonus the young leaves display bronzy hues.

The genus *Cladrastis* is barely known in cultivation, even the American species *C. kentuckea* remains a connoisseur's plant, so it is little wonder that the Chinese tree *C. sinensis* is a still rarer species. More is the pity as this small to medium-sized tree has a great deal to commend it. Late into growth it produces pinnate leaves and in July, on established trees, abundant panicles of white or pinkish flowers that are delicately scented. Though Wilson once again introduced the tree from western Hubei on his first trip, it has a wider distribution. I have seen it in the forests north of Kangding in western Sichuan and George Forrest collected it in Yunnan. An allied species was named for Wilson by the Japanese botanist Takeda though, sadly, I have never seen *C. wilsonii*.

Davidia involucrata remains one of Wilson's key introductions not only because it launched his career but because it is a tree of singular beauty, a big specimen when in full flower and as striking as any tree, be it temperate or tropical. The story of Wilson's quest for the dove tree and the subsequent tribulations he endured has been related so many times as to be not worth the repetition. We are fortunate indeed that his efforts were successful and we can enjoy old specimens such as the tree at Frensham Hall in Surrey, which rarely fails to be anything other than a stunning spectacle when unfolding its myriad bracts in May.

The Chinese tulip tree, *Liriodendron chinense*, is another species of abiding interest. Not only does it demonstrate the fascinating floristic link that China enjoys with the eastern United States (the only other member of this genus is the Appalachian yellow poplar *L. tulipifera*) but it is a tree of ancient lineage, a component of the pre-glacial flora of the Chinese forests. Add to this its great refinement and beauty – mature trees have a wonderful statuesque appearance –

and you have a winning combination. The uniqueness of the genus is emphasised by its leaves, which have two side lobes and a scalloped tip. In the Chinese tulip tree the newly emerging leaves are bronzy coloured and retain a silvery sheen on the underside, important characteristics that distinguish it from its closely related American cousin.

The forests of China are replete with climbing plants in a manner reminiscent of the tropical forests; they festoon the trees and alternately flaunt their flowers and fruit. So many are highly ornamental that it seems invidious to choose just one, however, I've always had a soft spot for *Lonicera tragophylla*. This yellow flowered species is quite different from other honeysuckles, its flowers are large and of the cleanest of colours and it lacks any spotting or flecking. Sadly, it also lacks scent: a decided disadvantage amongst a group of climbers renowned for their perfume but one that doesn't diminish it in my eyes.

The poplars are a ubiquitous group, containing trees of great vigour but short duration and for many people they have unfortunate connotations, forever associated with the bleak landscape of industrial England. Others find little of intrinsic quality to commend them, regarding them as dowdy or worse. And yet there are exceptions and one such is a poplar of great quality that deserves to be better known. Fittingly it commemorates the great man himself: *Populus wilsonii*. Sea-green, heart-shaped leaves are held with purpose from stout shoots, in time it forms a pyramidal canopy within which the leaves tremble in the best tradition of the genus. Sadly, a difficulty in propagating the tree and a lack of awareness conspire to keep it as a denizen of the discerning arboretum.

Finally, from God's own genus a rhododendron of quite exquisite beauty. *Rhododendron orbiculare* draws its specific epithet from the shape of its leaves, which immediately distinguish it within the genus. When grown in good light (a woodland edge is ideal) they are clean and crisp with deep, cordate bases of a bright jade green. In such a situation the plant will remain compact, retaining its skirt to the floor. In the best forms the flowers are held well above the foliage and are widely campanulate and of a clear pink, unsullied by any underlying colour.

Of the two species of tulip tree the North American *Liriodendron tulipifera* remains the commonest in cultivation. Few people know that a second species *Liriodendron chinense* is also native to China and although widespread it is also an uncommon tree in the wild. In virtually all such cases, and there are many examples, the Chinese species is a refined and distinguished tree much the superior of its North American counterpart.

Rhododendrons and China are synonymous and Wilson was a prolific collector of these signature plants. Of his many introductions *Rhododendron orbiculare* is amongst the best. Despite having a limited distribution it is variable and the best forms have clean, crisp leaves and beautiful clear pink flowers. When grown in good light it forms tight rounded specimens.

The fallen leaves of *Populus wilsonii*, a poplar of refinement and quality that deserves to be more widely grown and appreciated.

Photograph © Mark Flanagan

TONY KIRKHAM'S SELECTION

My personal top ten begins with *Emmenopterys henryi*, a member of the coffee family, Rubiaceae, because of the mystery and intrigue that surrounds its reticence to flower in cultivation. Wilson, who gathered it near to Ichang in Hubei province in 1907, described it as, 'one of the most strikingly beautiful trees of Chinese forests'.[9] On 2 October 1996 Mark and I were driving through Youyang County in Sichuan following a spell of collecting in the Micang Shan when we suddenly spotted a large specimen tree about 10 metres high and covered in fruit growing in a local farmer's garden. We shouted for the driver to stop and reverse back, to confirm that it was this elusive tree. It was covered in ripe, brown oval-shaped fruits with the brown bracts still intact, resembling those of *Schizophragma*. The farmer and his family kindly allowed us to collect the seed, even offering us the use of their ladder to access the upper canopy. After the excitement and thrill of finding Wilson's special tree we paid our thanks with a Kew guide book and continued on our journey to catch up with the rest of the team. This plant has so rarely been seen in flower outside China, especially in Europe, and flowered for the first time in Villa Taranto on Lake Maggiore, Italy in 1971. Imagine my delight when, in July 2004, I received a call from Bill McNamara saying that a specimen in Quarryhill Botanical Garden, California, grown from the seed we collected from that farmer's garden, had flowered. I was so elated that I immediately booked a flight to San Francisco to see and photograph this remarkable tree. I wasn't disappointed.

One of Wilson's favourite magnolias was *Magnolia parviflora*, now known as *M. sieboldii*, which he collected from the Diamond Mountains, Korea in 1918. He wrote, 'not even in the richest parts of China or Japan have I seen such extensive displays of pure pink and white as on the Diamond Mountains, where *Rhododendron schlippenbachii* and *Magnolia parviflora* dominate the undergrowth for miles and bloom to perfection.'[10] My favourite with a similar

China abounds with enigmatic plants, seemingly in harmony with the inscrutability of the Chinese race. *Emmenopterys henryi* is one of the most unfathomable of all – why does it resolutely refuse to flower with any regularity under British conditions?

Of the many plants named after Wilson his magnolia, *M. wilsonii*, is one of the most beautiful and a plant in full flower never fails to arrest the attention.

Wilson's eye for a good plant and his tenacity in introducing them is perfectly illustrated with *Meliosma veitchiorum*. This striking pinnate-leaved tree owes its presence in cultivation to Wilson's efforts and he was a strong advocate of its virtues.
Photograph © Harvey Stephens

flowering habit has to be *M. wilsonii*, named after the man himself and a native of woods in the provinces of western Sichuan, Yunnan and Guizhou. His first introduction in 1904 from the woods around Kangding never reached our gardens and it was from a Wilson collection four years later, in 1908, that the Arnold Arboretum in Boston sent plants to Léon Chenault, which were later introduced to the British Isles. This choice garden plant is best grown in a shady woodland situation in rich loamy soil and usually makes a multi-stemmed tree or large shrub to 7 metres tall. It is important to allow the plant to make height in its early years, as the fragrant, pendant, cup-shaped flowers consisting of 9 white tepals and a ring of bright crimson stamens are best viewed from beneath in May and June.

Meliosma veitchiorum is, in my opinion, one of the most unusual, striking, architectural plants growing in the temperate forests of western China and is immediately identifiable by its very stout, upright branching habit. I remember an original Wilson specimen growing by the old 'T' range at Kew when I was student and I always marvelled at its unusual form with its large pinnate leaves up to 80 cm long and bright red petioles, which become more apparent during the autumn months when the leaflets turn a rich yellow. The flowers are large-creamy white panicles around 50 cm long and bear purple fruits with approximately 1.5 cm diameter. Mark and I have seen this tree growing in the forests of western Sichuan on many occasions and we are always lost for words when we first see it. We collected seed in 1996 from a 15-metre high tree in the Micang Shan after a long and tireless search for several days of many trees which seemed to be shy at showing their harvest, eventually finding a specimen in a valley bottom by the edge of a river with fruits hidden at the very tips of the branches by the large leaves. It is a difficult tree to establish but we now have plants growing successfully in the arboretum at Kew and in the Savill Garden at Windsor.

Another uncommon tree found mainly in botanic collections is the monotypic genus *Poliothyrsis sinensis*, which joins other Chinese genera such as *Carrierea* and *Idesia* in the family Flacourtiaceae. This largely subtropical family has precious few warm temperate taxa in the mountains of western and central China. Augustine Henry found *Poliothyrsis sinensis* growing in the forests of western Sichuan through to western Hubei in 1889 but it wasn't until 1908 that Wilson introduced it to western gardens. It grows slowly, reaching 12 metres in height when mature. It needs a hot summer to produce flowers but when it does it is a spectacular sight. There is an original Wilson plant in the Berberis Dell at Kew under an Arnold Arboretum code, (AARB).

I have a fondness for lilacs but none more than *Syringa reflexa* and no one better describes it than Wilson in chapter 12, 'In Lilacdom' of *Aristocrats of the Garden*:

> *The most distinct of all lilacs is the new S. reflexa with narrow or broad flower clusters from nine to 12 inches long, suberect, nodding or pendent and sometimes hang downward like the inflorescence of the Wisteria. The expanding flower buds are bright red and the open flowers are pale rose colour.*[11]

The lilac is native to the forests of western Hubei where Wilson discovered it in 1901. Mark and I have seen and collected it growing on the Erlang Shan in western Sichuan.

Worthy of any winter border is the early spring flowering shrub *Corylopsis sinensis* with its abundant, beautifully-scented, lemon-yellow pendent racemes showing before the leaves. Whenever I think of this plant, I am always reminded of a night in a Chinese hotel room, where we had left various fruits out to dry before bagging the seed once it had shed. Halfway through the night there was a continuing barrage of objects firing across the floor and neither of us wanted to investigate in case it was something we didn't want to see. In the morning we woke to find that the hard rounded capsules

One of the unique features of the Chinese flora is the incidence of hardy members of otherwise subtropical families. The Flacouritaceae is a case in point; this mainly tender, evergreen family provides us with hardy, deciduous members such as *Poliothyrsis sinensis*.

Lilacs enjoy an undeserved reputation as being stolid and unremarkable garden plants – showy and scented in flower, dowdy and workaday otherwise. *Syringa reflexa* gives lie to this reputation, producing a large, arching shrub with big oval leaves and masses of pendulous richly coloured flowers.

which make up the fruits of *Corylopsis* had opened with the overnight heating and the small, shiny black seeds had been propelled in all directions to every corner of the room. We spent a good part of the morning on our hands and knees searching the floor for the rest of the seed.

It is difficult to choose one climber of Wilson's, but I think *Parthenocissus henryana* has to be a good choice for a north-facing wall to get the best out of it. Discovered by Augustine Henry in 1885 but introduced by Wilson from his first expedition in 1900, it is a beautiful self-clinging species; but one that often requires some additional support. During the summer months when grown out of direct sunlight it has green or bronze leaflets with silvery-white inter-veined variegation. In the autumn it turns a magnificent bright red that brings to life any dark and uninteresting wall.

Ethereal is a good word to describe the qualities of the strictly Asian genus *Corylopsis*. In the reduced spring light of the woodland garden, where these plants are most at home, *C. sinensis* has a ghost-like appearance that never fails to intrigue.

The Virginia creeper and Boston ivy are surprisingly common climbing plants belonging to the genus *Parthenocissus*. By contrast, the far less well-known *P. henryana*, a Wilson introduction, has qualities the other two lack. The silver veinal variegation is striking and the autumn colour much more complex.

Often seen as a border filler in urban parks and squares, it is unjust to lump the magnificent evergreen shrub *Viburnum rhytidophyllum* with the likes of cherry laurel and the spotted laurel: it should be given a more defined and purposeful place in the border. It produces large, dark-green, tough leathery corrugated leaves almost resembling those of a rhododendron, which make a good back cloth for the abundant creamy-white flowers produced during May. This plant was crossed with *V. lantana* to create the larger, more vigorous hybrid *V. ×rhytidophylloides*.

Possibly one of the most popular and best value garden plants in British gardens that Wilson introduced from China is the Chinese dogwood, *Cornus kousa* var. *chinensis*. This tree or large shrub flowers in June and must be one of the most floriferous of shrubs with the longest lasting flowers, the bracts so virginal-white that a single specimen or a group growing in a garden resemble a column or mound of snow. Following the flowers are the matt red strawberry-like fruits and red autumn colour; an all round winner in the garden.

Stunning flowers, masses of red strawberry-like fruits and an attractive piebald bark make the Chinese flowering dogwood a stellar plant. No surprise then that this is a celebrated Wilson introduction.

Unassuming evergreens have a key role in the garden and landscape. They are generally used to provide shelter or screening but most tend to be drab. *Viburnum rhytidophyllum* is an evergreen of distinction. Dark green corrugated leaves, felty on the underside, are densely set to produce a handsome and eye-catching shrub which is equally at home on acid or alkaline soils.
Photograph © Mark Flanagan

Rhododendron argyrophyllum is a plant with an easy disposition and propensity to flower well every year. Add to this its neat foliage with a silvery underside and you have a winning combination. A star performer in many of our best known rhododendron gardens. This plant is growing at Dawyck Botanic Garden under WILS 1210.

Finally, like Mark, I have to choose a rhododendron from the many that Wilson introduced to our gardens: *Rhododendron argyrophyllum*, a graceful large leafy shrub with large, loose heads of pink flowers in May. An incredible sight that once seen you will always remember is the large group of plants under Wilson 1210 in full flower at the top of the garden at Dawyck, one of the satellite gardens of the Royal Botanic Garden Edinburgh, on a bright spring early evening. This plant epitomises for me the floral treasure house that is western China and the keen eye of the man who did so much to collect and popularise these gems.

The easier to propagate and more utilitarian of Wilson's collections, largely those in my first group listed on p. 215 were quickly adopted by the nursery trade and produced, over the years, by the million. They are forever secure in cultivation, though their links back to Wilson have long been lost as it is difficult, if not impossible, to connect any such plants directly back to original collection numbers. We must, therefore, constantly remember that we owe Wilson a debt for their presence in our gardens. The final group, our top twenty, are quite different. As suggested, most are only found in a very few collections and many are still the original plants raised from Wilson's seed; they provide a direct link back to him and the lost world in which they originated. Some have been propagated by their host gardens and distributed. Some years ago Kew micropropagated its Wilson collection of *Emmenopterys henryi* and distributed it widely with the result that not only do we have young plants but it is now found in many more gardens. But this was, by and large, an isolated event and many of these 'Wilson originals' remain as a precious few, or indeed single, specimens in our most specialist gardens.

With our growing awareness of Wilson and the research we conducted into his collecting activities, Tony and I looked at the dwindling band of authentic Wilson plants with a degree of alarm. All are now entering their second century and we felt that some concerted effort should be made to arrest their decline. The workshop that we conducted for the PlantNetwork of Britain and Ireland (see p. 23) supported the idea that something should be done and most importantly some of the key players who could actually make this happen were represented at the workshop. David Knott, Curator of the Royal Botanic Garden Edinburgh, the collection with the best representation of Wilson originals, pledged his support as did several of the smaller collections. When combined with Kew's collections and the holdings of rhododendrons held in my own gardens in Windsor Great Park, we had more than a quorum to start the ball rolling.

In the winter of 2006–2007 I began in a small way to initiate a propagation programme. Scion wood from Wilson's birches and maples arrived from Kew and Edinburgh as did pieces of *Catalpa fargesii*, an ailing champion tree at Kew in need of rapid regeneration. Windsor's propagator, Shelley Richards, used her grafting skills to coax about 60 per cent of the material to form a successful union with their respective rootstocks and once lined out into nursery frames they formed strong-growing maiden trees. By the end of the first season most were returned back to Kew and Edinburgh with a few retained at Windsor. In the case of Kew the process was completed when Tony and I were present at the planting of a young sapling of *Catalpa fargesii* WILS 664 by our principal sponsor, Ian Bond and his wife Caroline, next to the original tree behind the Temperate House close to Holly Walk.

The task has been barely started but we hope to quietly work our way through as many of the Wilson originals as possible. It won't be easy as many of the plants are notoriously difficult to propagate, lack vigour and produce physiologically retarded shoots, making the process of initiating roots or callus a daunting job. Nonetheless, the rewards are high and with each success we respect the memory of the great Ernest Henry 'Chinese' Wilson as well as securing his living legacy for future generations.

Ian and Caroline Bond plant the first of the newly grafted trees *Catalpa fargesii* WILS 664 in autumn 2008 only a few metres from Wilson's original specimen in the arboretum at Kew.

Young grafted trees of Wilson's original collections provide the promise that the legacy of the great man can endure into the 21st century. L to r – *Tetradium daniellii*, *Malus prattii*, *Liquidambar formosana* var. *monticola* (two trees).

EPILOGUE

Inexorable, relentless, inevitable.

The sheer certainty of the event and its tragic consequences provide the numbing sense of pathos that surrounds the earthquake which shook Sichuan at 2.28pm local time during the afternoon of Monday 12 May 2008.

The forces involved, inexorable and relentless, made the earthquake an inevitability. This is an encounter that has a history stretching back for tens of millions of years. Once the Indian subcontinent collided with the Asian mainland 40 to 50 million years ago the clock began to tick towards the tragedy that unfolded. The forces involved are unimaginable. The collision created Mount Everest, lifting the seabed skywards by over 9 kilometres, and it continues to push the summit upward by 0.5 centimetres every year.

Other effects of the continuing movement are felt right across the mid-continental area of Asia from Afghanistan in the west to China in the east. The stresses formed have created an innumerable series of fault lines that continue to buckle under the pressure imposed by the northward movement of the Indian landmass. In the vicinity of central Sichuan the stress line is known as the Longmenshan fault and it is a particularly vulnerable part of the earth's crust. Here the crumpling and fracturing mountains to the west meet the hard, unyielding rocks that underlie the Sichuan Basin. A previous rupture of the Longmenshan fault caused the 1933 earthquake (described in Chapter 7).

In the intervening 75 years the stresses have been inexorably building. Seismologists now believe a catastrophic failure along the Longmenshan fault caused the fracture line to shear in opposite directions by up to 10 metres in two distinct phases,[1] with a massive release of energy measuring 7.9 on the Richter scale. The United States Geological Survey rates this as the fourth most powerful earthquake in China in the last 600 years.[2]

The consequences of the cataclysm have been well-reported in the media, with harrowing on the spot reports, eyewitness testimony and traumatic film footage of the aftermath. Tony and I heard about the earthquake at almost the same time. We feared the worst: knowing well the precipitous, unstable river valleys of this area of Sichuan, it took little imagination to appreciate what might be the results of such a catastrophe. Many of the towns and villages are hemmed in by steep mountainsides, some rising for hundreds of metres above the settlements. As news filtered out, it also became clear that we had actually travelled through the worst affected areas in 2006, during our quest to retrace Wilson's 1910 journey to Songpan (as detailed in Chapter 6).

More disconcertingly, we had spent the night in Beichuan, the town hit hardest by the

A poignant reminder of that fateful day in 2008 when the earthequake rocked the lower Min Valley.

earthquake. The very hotel we had stayed in was apparently reduced to rubble, many of the unsuspecting guests entombed in a mass of concrete, tangled girders and splintered glass. Of all the horrors visited on this quiet county town none was more terrible than the fate of its middle school, where 1,000 children lost their lives as a mass of tumbling rocks engulfed the school buildings.

Not surprisingly, a range of emotions arose within me as I began to digest the scale of this natural disaster. My first concern was for my closest friends in China – Yin Kaipu, Zhong Shengxian and Wang Hangming – although they live to the south of the disaster area, in Chengdu, early reports suggested that the city had not been immune to the effects of the earthquake. This concern was heightened when Tony told me that he had received no response to his numerous emails; Zhong was normally so prompt to reply. Eventually our worst fears were allayed; all three were safe. But not so several of their colleagues, most of whom had been engaged in fieldwork in the mountainous areas close to the epicentre of the earthquake.

The scale of the damage also suggested that many of the scenes we had photographed in the area during our 2006 visit had been irrevocably altered. Had the bridge at Beichuan (see p. 149) survived? What about the nearby villages such as Kaiping and Xiao Ba (pp. 156–160)? Furthermore, Zhong had an unsubstantiated report that Mr and Mrs Liang, with whom we had established a direct link back to Wilson's visit, and their grandchildren had been killed in the village of Kaiping.

Slowly, Tony and I began to feel that we had some unfinished business, that our narrative had a missing chapter and that several of our important photographs needed to be updated. Though the manuscript and images were rapidly moving towards publication we felt the book was now incomplete. By the late summer of 2008 we were agreed that we had to go back. We broached the question of a return visit with Yin and Zhong. Not surprisingly, their response was guarded. The situation in the earthquake zone was still chaotic, the massive amounts of debris had blocked several of the rivers and, as in 1933, huge 'quake lakes' had been created. Everywhere, roads were still impassable, accommodation impossible to find and the spectre of disease ever present. Nonetheless a return visit was deemed to be possible. We continued negotiating into the autumn until it was finally agreed that we could return to China at the end of October.

Nothing could have prepared us for what we were to witness.

On 2 November 2008 we stood on the hillside overlooking the stricken town of Beichuan; a tortuous and difficult journey had taken us this far. At Liang Feng Ya the road was sealed off, the authorities determining that it was too dangerous to allow anyone other than officially sanctioned personnel to continue further. We covered the last section on foot. The walk to the hillside had a surreal element. At this most tragic and sensitive of locations hawkers had been allowed to trade their wares. Pre and post earthquake photographs of Beichuan were for sale, as were photographs of the visits to the site by Wen Jianbiao and Hu Jintao, the prime minister and president of China.

Tony and I continued on heading up to the hill with the procession of people. The mood became increasingly solemn as we got closer to the viewpoint. A fence topped by razor wire prevented access beyond the approved boundary and attentive police officers stood guard. Finally we took up our positions high above the valley of the Tianjian River within which the town of Beichuan lay. It took time to understand the scene. As the details became clear I spotted the Beichuan Hotel, our overnight domicile on our previous visit. Somewhat surprisingly to me, it was not a heap of rubble; it almost seemed intact. However, as I squinted into the late morning sunshine I could see that the bottom floors had collapsed and that the rest of the building was clearly unstable.

Further into the town the devastation was more complete; a huge landslide from the hillsides to the right had buried most of the east section and in the centre nothing but rubble remained.

If this wasn't awful enough the whole of the former entrance to the town was awash with suffocating mud: a consequence of a torrential downpour on 23rd September 2008. Over 450 mm of rain was dumped on the valley in a 12-hour period, causing mudslides to cascade down the mountainsides. How cruel and fickle nature can be.

Tony stood ahead of me taking images of the Armageddon below. Despite the crush of people a respectful silence was maintained, the only sound was people weeping. Zhong recounted the devastating statistics to us. Prior to the earthquake 30,000 people had resided in Beichuan, unofficially it is now believed that two-thirds had been killed or buried alive. More poignantly, Beichuan is the principal residence of the Qiang people, one of China's minority races, who have lived in these valleys for countless generations. Their exact numbers are not known with any degree of accuracy but they have always been a close-knit and small community and for this catastrophe to befall them in their heartland is a shattering blow.

The entire town of Beichuan is now closed to visitors by a razor wire fence. The Beichuan Hotel where we lodged in 2006 is the red-roofed building in the centre of the image.

We slowly returned to our vehicle, the stalls of the hawkers and the gambolling of local children seemingly at odds with the unimaginable horror below us. Perhaps though, this incongruity attests to the resilience of the human spirit and the age-old maxim that life goes on. Ultimately, the decision of the Chinese authorities to allow Beichuan to remain largely untouched and for it to become a giant mausoleum is perhaps the most respectful way to honour the dead. A new town will be built but not on the site of the old one.[3]

The visit to Beichuan was the culmination of our return trip, a journey that had been both harrowing and uplifting. In the space of just five or six days Tony and I had undergone a rollercoaster journey both in terms of the speed of the trip and the emotions we had experienced. Our team was once again led by Dr Yin Kaipu, the retired ecologist from the Chengdu Institute of Biology, Zhong Shengxian, the Head of Library Services at the Institute and our capable driver and general factotum Wang Hangming.

Our first target was the bridge between Beichuan and Yuli: had it survived the devastation? The drive up the road leading north along the Min Valley to Yuli was a desperate struggle; town after town was a scene of carnage. At the southern end of the valley the tourist city of Dujiangyan, though not a ruin, had sustained irreparable damage to 80% of its buildings, making most of them uninhabitable and forcing the residents to take refuge in the huge temporary camps set up by the local authorities. Fortunately Li Bing's age-old engineering works (described in Chapter 5) were unaffected.

The earthquake left the lower Min Valley desolated.

As we progressed the scene became ever more dramatic. The middle parts of the valley had changed totally. In many places the mountainsides were devoid of vegetation and consisted of huge screes from high on the precipitous slopes right down to the river itself. The road on the opposite, west, bank had more or less been completely obliterated and it seemed as if little effort had been made to reinstate it. Perhaps it would never be used again. A further concern was for the habitat itself as this is the home of the regal lily, as described previously. How had this plant fared given that it was endemic to the Min Valley? Though abundant in this location surely few of the bulbs could emerge from beneath the recently created screes?

Our journey of 50 kilometres became a five-hour ordeal, the result of traffic chaos as drivers tried to negotiate the shattered carriageway, broken bridges and huge rock falls. A 'one day, one-way' system had been established but this still didn't prevent vehicles trying to make their way south, which only added to the pandemonium. At the town of Yingxiu we arrived at the supposed epicentre of the earthquake. Once again the local school had been destroyed and much of the housing was reduced to rubble. The temporary encampments provided a cleaner and more ordered environment amongst the general disorder.

At various points the traffic was held up as certain sections of the road, where huge boulders continued to fall from the slopes above, were deemed too dangerous to allow stationary vehicles to wait. One such place was the village of Luo Quan Wan where Tony and I stopped to speak to local people. One old man told us that on the afternoon of 12

Despite the one-way traffic system set up in the Min Valley, allowing vehicles to negotiate the roads and bridges destroyed by the earthquake, heavy traffic brought the roads to a standstill for several hours at a time.

The last 100 metre drive into the small village of Luo Quan Wan, surrounded by mountainsides left devoid of vegetation, was silent and eerie.

The expressions on the faces of the residents of Luo Quan Wan say it all.

May he had been outside his house on the roadside when the earthquake began. He was thrown to the ground and forced to remain prostrate for three minutes as all around him the air thundered with falling rocks and shattering masonry. When the tremors subsided the air was so thick with dust that he was unable to see his hand in front of his face. His home had been destroyed and he pointed to a rock fall on the opposite side of the road where several other houses had stood. All the occupants, numbering 30 people, had been buried and their bodies had not been recovered. One of the bereaved women also came to speak to us; she had lost her grandson, granddaughter and daughter in law. All the residents of this ill-starred place had a haunted expression that told of their pain and sorrow.

The following day we left the Min Valley and headed over the mountains along the Gan Gu River. The devastation was less here and on consulting the map we realised we were temporarily in the lee of the fault line. Soon, however, the telltale signs of the catastrophe reappeared along with the determined efforts of the local people who were engaged on a huge rebuilding programme. Xiao Zhong told us that the Chinese government is contributing 20,000 yuan (about £2,000) to each family made homeless by the earthquake, with a further 5,000 yuan for each family member who was killed. In addition, every Chinese province has adopted a local county within the earthquake zone to provide financial and material support. Eventually we emerged from the mountains at the town of Yuli, Wilson's Shih'chuan Hsien. Immediately we were confronted by a large body of water where previously dry land had existed. This was the notorious Tianjian 'quake lake' which resulted from the blocking of the Tianjian River further down the valley.

We entered the village with some trepidation and were greeted by a scene of utter devastation. It was interesting to note that the modern concrete structures had sustained the most damage; the older wooden buildings were still largely extant. Once again discussions with the local people revealed an incredible story. An old lady, Liu Shu Fen, told us that the main street where we were standing

In every village around the earthquake zone, local people were busy re-building their homes using locally found building materials and traditional Chinese techniques.

The Tianjian 'quake lake' at Yuli.

The authors observing the sheer destruction in the town of Yuli, resembling a war zone.

Liu Shu Fen relaying her account of the earthquake to the team in Yuli.

was flooded to the height of the rooftops by the quake lake which, with the assistance of the Chinese Army, was gradually drained over a seven-week period to allow the residents to return. She also told us that the lake had completely submerged the bridge we were heading for. This was disappointing news. Apparently the water was so deep it rose almost to the tops of the adjacent hillocks, which are a key part of the scene in Wilson's 1910 image and our corresponding view from 2006. There is little likelihood that this area will ever again be accessible by anything other than a boat.

We had no alternative but to turn north towards Kaiping, the home village of Mr and Mrs Liang. This was likely to prove to be the most harrowing part of the whole trip. On our previous visit we had been delighted to discover that Liang Xue-Fu was a great-grandson of the widow whose shrine Wilson had admired when he visited Kaiping in August 1910. Now there was a good chance that he and several of his family members had been victims of the earthquake. As we entered the village almost the first person we saw was Mr Liang, looking perfectly hale and hearty. We were invited into his house where his equally robust wife greeted us. Over lunch Mr Liang recounted his experiences. He had been on the mountainside during the quake but had emerged unscathed and his wife had been in the relative safety of their old wooden home. Indeed, no one in Kaiping had been killed but the village had been cut off from the outside world for six days as road, telephone and mobile phone links had been severed. This had been their main anxiety, as their two grown up children were living in Beichuan. Fortunately as soon as communications were re-established they were able to confirm that all was well. We took our leave of Mr and Mrs Liang, happy in the knowledge that this couple will enjoy many more years together.

Sadly this happy outcome was not to be the case when we picked up our journey in the Min Valley again. At Pai Shan Yin we are greeted by the mother of Xi Xiao Long, the little boy we had come to know from previous visits to this village (as described in Chapter 7). It was clear that all was not well. The boy and his younger sister were at school but we soon learned that her husband, the children's father, had been killed during the earthquake. He was 32 years of age. The memory was still very vivid for her and the emotional scars raw and painful; tears cascaded down her face as she spoke to Xiao Zhong. Her area – Maoxian County – is being supported by Shaanxi Province and a wealthy man from Shandong Province is also providing financial assistance to the family so her future seems reasonably secure, though this provides little compensation for the loss of her husband.

Mr and Mrs Liang, thankfully safe outside their home in Kai-Ping.

Looking up the side valleys of Dan Yun Xia towards the mountain range Xuebaoshan, flanked with the autumnal tints from the temperate forest.

We had one final task to undertake which provided an incredible conclusion to our visit. Apparently, the long lost outdoor theatre at Mianzhu, which Wilson mistook for a temple, had been completely rebuilt as a result of the interest shown in it by ourselves and Dr Yin. I found this difficult to conceive and it certainly made the journey back south worthwhile. The fastest route was via Songpan and Pingwu down the valley of the Fujian River. This allowed us another visit to Xuebaoshan Jiangzi, the pass from where the Snow Treasure Peak can be seen. Our previous visits to this site had all been much earlier in the year and we much looked forward to seeing the mountain and surrounding tops in their early winter garb. Sadly, we were disappointed, as cloud cover obscured the main peak – though the views still proved to be breathtaking.

The real treat emerged as we drove down from the pass. Beyond the entrance to the World Heritage Site at Huanglong we entered the Dan Yun Xia. This majestic valley was impressive in midsummer of 2006 with its fine array of broadleaved and coniferous trees; but in the late autumn it surpassed our highest expectations. The first trees to catch the eye were *Larix potaninii,* which had turned a beautiful clear yellow made all the more effective by the dark foliage of the surrounding silver firs and spruces. Further down, maples, limes, hornbeams and rowans joined the spectacle. The orange tones of *Acer davidii* were exceptional and quite unlike anything seen in cultivation, even on the best trees. In the understorey, *Viburnum betulifolium* and *Berberis wilsoniae* were weighed down with huge crops of fruits. Finally and fittingly, Wilson's own maple, *Acer sinense* subsp. *wilsonii,* flaunted its dark red autumn leaves.

The majestic valley Dan Yun Xia at Huanglong Reserve, with the yellow autumn tints of *Larix potaninii* against the dark foliage of spruce and silver firs.

(*far left, top and middle*) *Acer davidii* in its various guises of autumn colour, reds and yellows.

(*far left, bottom*) *Viburnum betulifolium* with an abundance of translucent red berries that will last well into the spring.

(*left*) Mrs Wilson's Barberry, *Berberis wilsoniae*, which Wilson introduced in 1904 and named after his wife.

The beautiful autumn leaves of *Acer sinense* subsp. *wilsonii*, which was introduced into cultivation by Wilson in 1907.

0250. "A study in architecture". Village temple at Kung-ching-chiang. Near Hien chu Hsien. Alt. 1900 ft. Aug. 10, 1910.

'"A study in architecture".
Village temple at Kung-ching-
chiang. Near Hien chu Hsien.
Alt. 1,900 ft. Aug. 10, 1910.'

The newly constructed theatre in the city of Mianzhu.

Within a few hours we arrived at the city of Mianzhu. Yin and Wang had visited the rebuilt playhouse earlier in the year but had trouble re-finding it. Eventually we discovered it tucked away amongst the surrounding nondescript buildings. And what a sight it proved to be! It was much bigger than I had imagined and the craftsmanship and attention to detail of the work was impressive, recalling the best traditions of Imperial China, all of this garnered from Wilson's photograph. The building sat at the head of a reconstructed historic Chinese street, part of a tourist initiative. The local museum curator, Ning Zhiqi, had persuaded the town's distillery to underwrite the cost of the work. Sadly, the pace of the scheme had slackened due to resources being directed to earthquake recovery. The new

The roof's fine ornate detail. Some of the tips have been damaged by the earthquake.

Craftsman busy at work on the finer details of the stage.

building itself had, ironically, sustained some limited damage. However, we still found a great many craftsmen hard at work, all of whom were more than happy for Tony to take photographs and for us both to climb onto the stage from the rear area. All joined in the laughter as Tony exercised his tonsils in a mock rendition of Chinese opera.

This exciting development provided a final high note to our visit and balanced the otherwise difficult elements of the journey. For the future, however, it indicated the level of interest the work and memory of E. H. Wilson continues to exert in China. Yin Kaipu is taking our understanding of Wilson's journeys to new levels, having now matched several hundred of his Chinese images in both Sichuan and Hubei.

Tony shows his excitement at finding the new theatre by giving a rendition of a Chinese opera on stage to the local onlookers and workers.

0356. My Chinese collectors, all faithful and true. Ichang. Feb. 15, 1911.

Whilst in Yichang, the old treaty port on the Yangtze River and Wilson's first base in China, Yin placed an advert in several local papers accompanied by Wilson's photograph of his 'Chinese collectors', all of whom had been recruited locally. Incredibly two men came forward, the great-great-grandsons of two of the people in the photograph, including the person we believe was Wilson's head man. At first they were reticent to present themselves but out of respect for their great-great-grandfathers had decided to meet Dr Yin. Wilson's photograph had for a long time adorned both family homes but it was destroyed by the Red Guards during the Cultural Revolution. Sadly, neither of the men – Yin Hong and Wang Hongbing – knew the full name of their ancestor. We can at least be sure of the family name of two of Wilson's team, including the man (fourth from the right in the photograph) who appears to have been his trusty lieutenant, as he occupies the most prominent position within the group and also appears in several other photographs by Wilson. We now know that this man was Mr Yin, and that the man standing behind him and to his left is Mr Wang. With this revelation Tony and I felt we had drawn a satisfactory line under our quest to find E. H. Wilson, and that our work could now be presented to a wider audience in the hope that they too would be captivated by this most enigmatic of Edwardian adventurers.

'My Chinese collectors, all faithful and true. Ichang. Feb 15, 1911.' This is one of the last images Wilson took before his departure from China in February 1911.

REFERENCES

INTRODUCTION

1. Archives of the Arnold Arboretum. Sargent correspondence. 28 January, 1910.
2. Clausen, K. S. and Hu, S–Y. (1980). Mapping the collecting localities of E. H. Wilson in China. *Arnoldia* 40 (3) 139–145.
3. Flanagan, M. and Kirkham, T. (2005). *Plants from the Edge of the World*. Timber Press. Portland, Oregon.
4. Wilson, E. H. (1913). *A Naturalist in Western China*. Methuen and Co. London. Vol. 1. p. 205.
5. Flanagan, M. (2002). Wilson's lost tree. *Arnoldia* 61 (3) 12–17.
6. Flanagan, M. and Kirkham, T. (2005) *op. cit.*
7. Wilson, E. H. (1913) *op. cit.* Vol. 1. p. 131.
8. Rae, D. et al. (2006). *Catalogue of Plants*. Royal Botanic Gardens, Edinburgh. p. xxi.
9. Bean, W. J. (1976). *Trees and Shrubs Hardy in the British Isles*. Vol. 1. 8th edn. John Murray. London. p. 15.
10. Wilson, E. H. (1913). *op. cit.* Vol. 2. p. 4.
11. Schaller, G. B. (1993). *The Last Panda*. The University of Chicago Press. Chicago. pp. 148–149.
12. Del Tredici, P. 'Another seedy Watergate story'. *New York Times*. 17 February 1972.

CHAPTER ONE: ERNEST HENRY

1. Farrington, E. I. (1931). *Ernest H Wilson, Plant Hunter*. The Stratford Company Boston, Massachusetts.

    Briggs, R. W. (1993). *'Chinese Wilson'. A Life of Ernest H Wilson 1976–1930*. HMSO. London.
2. Huntford, R. (1979). *Scott and Amundsen*. Hodder and Stoughton. London.
3. McLean, B. (2004). *George Forrest Plant Hunter*. Antique Collectors' Club. Woodbridge, Suffolk.
4. Cox, K. ed. (2001). *Frank Kingdon Ward's Riddle of the Tsangpo Gorges*. Antique Collectors' Club. Woodbridge, Suffolk.
5. Shulman, N. (2002). *A Rage for Rock Gardening*. Short Books. London.
6. Farrington, E. I. (1931) *op. cit.*
7. Wilson, E. H. (1913). *A Naturalist in Western China*. Methuen and Co. London. Vol. 1. p. 107.
8. Veitch, J. H. (1906). *Hortus Veitchii*. J Veitch and Sons. London. p. 96.

9. Shipley Cunningham, I. (1984). *Frank N Meyer. Plant Hunter in Asia*. Iowa State University Press. Ames, Iowa. p. 60.

10. Wilson, E. H. (1913). *op. cit.* Vol. 1 p. 85.

11. Wilson, E. H. (1913). *op. cit.* Vol. 1 p. 148.

12. James, L. (1995). *The Rise and Fall of the British Empire*. Abacus. London. pp. 169–349.

13. Cox, K. ed. (2001). *op. cit.* p. 10.

14. Wilson, E. H. (1913). *op. cit.* Vol. 1 p. 135.

15. Archives of the Arnold Arboretum. Wilson Field Journal. Fourth Expedition. 3 September 1910.

CHAPTER TWO: WAWU – LUDING, DADU – KANGDING

1. Wilson, E. H. (1913). *A Naturalist in Western China*. Methuen and Co. London. Vol. 1. p. 23.

2. Wilson, E. H. (1913). *op. cit.* Vol. 1 p. 244.

3. Wilson, E. H. (1913). *op. cit.* Vol. 1 p. 231.

4. Wilson, E. H. (1913). *op. cit.* Vol. 1 p. 234.

5. Wilson, E. H. (1913). *op. cit.* Vol. 1 p. 236.

6. Wilson, E. H. (1913). *op. cit.* Vol. 1 p. 234.

7. Wilson, E. H. (1913). *op. cit.* Vol. 2 p. 92.

8. Xiong Lei (2003) Blazing the tea-horse trail. *China Daily*. 13 June 2003.

9. Chang, J. and Halliday, J. (2006). *Mao. The Unknown Story*. Vintage Books. London. pp. 158–160.

10. Wilson, E. H. (1913). *op. cit.* Vol. 1 p. 197.

11. Wilson, E. H. (1913). *op. cit.* Vol. 1 p. 205.

CHAPTER THREE: MYSTERY TOWERS OF DANBA

1. Wilson, E. H. (1906). Leaves from my Chinese notebook. *Gardeners' Chronicle* Vol. 39. 17 February. p. 101.

2. Wilson, E. H. (1906). *op. cit.* 24 March. p. 179.

3. 'Lonely Planet writer vanishes on trek in Tibet bandit country'. *The Times*. 5 June 2007.

4. Goodman, J. (2006). *Joseph F. Rock and His Shangri-La*. Caravan Press, Hong Kong. p. 97.

5. Wilson, E. H. (1906). *op. cit.* 3 March. p.138.

6. Hibbert, C. (1970). *The Dragon Wakes. China and the West 1793–1911*. Longman. London. p. 238.

7. Edgar, J. H. (1908). *The Marches of the Mantze*. London. China Inland Mission.

8. Wilson, E. H. (1906). *op. cit.* 3 March. p. 139.

9. Wilson, E. H. (1906). *op. cit.* 17 March. p. 166.

10. Wilson, E. H. (1906). *op. cit.* 24 March. p. 179.

11. Grey Wilson, C. (1993). *Poppies: the Poppy Family in the Wild and in Cultivation*. London. B T Batsford. p. 186.

12. Gardeners Chronicle Vol. 36. 1 October. 1904. p. 240.

13. Grey-Wilson, C. (1996). The Yellow Poppywort and its Allies. *The New Plantsman* 3 (1) 22–39.

14. Wilson, E. H. (1906). *op. cit.* 24 March. p. 179.

15. Wilson, E. H. (1906). *op. cit.* 17 March. p. 165.

16. Wilson, E. H. (1906). *op. cit.* 24 March. p. 180.

17. Wilson, E. H. (1913). *A Naturalist in Western China*. Methuen and Co. London. Vol. 1. p. 170.

18. Wilson, E. H. (1913). *op. cit.*

19. Wilson, E. H. (1913). *op. cit.* Vol. 1. p. 200.

20. Wilson, E. H. (1913). *op. cit.* Vol. 1. p.196.

21. Wilson, E. H. (1913). *op. cit.* Vol. 1. p. 195.

22. Wilson, E. H. (1913). *op. cit.* Vol. 1 p. 195.

23. Wilson, E. H. (1913). *op. cit.* Vol. 1. pp. 191–192.

24. Wilson, E. H. (1913). *op. cit.* Vol. 1. p. 161.

CHAPTER FOUR: THE DREADED PAN-LAN SHAN

1. Wilson, E. H. (1913). *A Naturalist in Western China*. Methuen and Co. London. Vol. 1 p. 189.
2. Wilson, E. H. (1913). *op. cit.*
3. Wilson, E. H. (1913). *op. cit.* Vol. 1. p. 184.
4. Hosie, A. (1905). *Parliamentary Report on a Journey to the Eastern Frontier of Thibet.* p. 69.
5. Wilson, E. H. (1913). *op. cit.* Vol. 1. p. 180.
6. Wilson, E. H. (1913). *op. cit..*
7. Wilson, E. H. (1913). *op. cit.* Vol. 1. p. 179.
8. Wilson, E. H. (1913). *op. cit.* Vol. 1. p. 177.
9. http://www.biodiversityhotspots.org/xp/hotspots/china/Pages/biodiversity.aspx.
10. Wilson, E. H. (1913). *op. cit.* Vol. 1. p. 176.
11. Wilson, E. H. (1913). *op. cit.* Vol. 1. p. 171.
12. Briggs, R. W. (1993). *'Chinese Wilson'. A Life of Ernest H Wilson 1976–1930.* HMSO. London. p. 53.

CHAPTER FIVE: DIG THE BED DEEP, KEEP THE BANKS LOW

1. Wilson, E. H. (1913). *A Naturalist in Western China*. Methuen and Co. London. Vol. 1. p. 108.
2. Wilson, E. H. (1913). *op. cit.* Vol. 1. p. 6.
3. Dujiangyan City Tourist's Manual. p. 8.
4. Wilson, E. H. (1913). *op. cit.* Vol. 1. p. 109.
5. Wilson, E. H. (1913). *op. cit.* Vol. 1. p. 105.
6. Wilson, E. H. (1913). *op. cit.* Vol. 1. pp. 105–107.
7. Hosie, A. (1905). *Parliamentary Report on a Journey to the Eastern Frontier of Thibet.* p. 76.
8. Wilson, E. H. (1913). *op. cit.* Vol. 1. p. 107.
9. Wilson, E. H. (1913). *op. cit.* Vol. 1. p. 171.

CHAPTER SIX: A MIDDLE WAY TO SONGPAN.

1. Sargent, C. S. (ed) (1913). *Plantae Wilsonianae*. Harvard University Press. Cambridge, Massachusetts. Vol. 1. p. vi.
2. Wilson, E. H. Unpublished manuscript 'Wilson plants in cultivation'. Archives of the Arnold Arboretum. Series W. XIII. Box 19, Folder 1.
3. Wilson, E. H. (1913). *A Naturalist in Western China*. Methuen and Co. London. Vol. 1. p. 116.
4. Chvany, P. J. (1976). E H Wilson as a photographer. *Arnoldia* 36 (5) 181–236.
5. Wilson, E. H. (1913). *op. cit.* Vol. 1. p. 118.
6. Wilson, E. H. (1913). *op. cit.* Vol. 1. p. 118.
7. Wilson, E. H. (1913). *op. cit.* Vol. 1. p. 119.
8. Wilson, E. H. (1913). *op. cit.* Vol. 1. p. 120.
9. Wilson, E. H. (1913). *op. cit.* Vol. 1. p. 119.
10. Hosie, A. (1905). *Parliamentary Report on a Journey to the Eastern Frontier of Thibet.* p. 10.
11. Pratt, A. E. (1892). *To the Snows of Tibet through China*. Longmans. London. p. 126.
12. Wilson, E. H. (1913). *op. cit.* Vol. 1. p. 121.
13. Shaughnessy, E. L. (ed). (2000). *China. The Land of the Heavenly Dragon*. Duncan Baird Publishers. London. p. 48.
14. Wilson, E. H. (1913). *op. cit.* Vol. 1. p. 121.
15. Wilson, E. H. (1913). *op. cit.* Vol. 1. p. 126.
16. Wilson, E. H. (1913). *op. cit.* Vol. 1. p. 131.
17. Wilson, E. H. (1913). *op. cit.* Vol. 1. p. 137.
18. Wilson, E. H. (1913). *op. cit.* Vol. 1. p. 139.
19. Wilson, E. H. (1913). *op. cit.* Vol. 1. p. 140.
20. Wilson, E. H. (1913). *op. cit.*

CHAPTER SEVEN: NEMESIS IN THE MIN VALLEY

1. Briggs, R. W. (1993). *'Chinese Wilson'. A Life of Ernest H Wilson 1976–1930*. HMSO. London. p. 37.
2. Wilson, E. H. (1906). Leaves from my Chinese notebook. *Gardeners' Chronicle*. Vol. 39. 23 June. p. 402.
3. Wilson, E. H. (1913). *A Naturalist in Western China*. Methuen and Co. London. Vol. 1. p. 144.
4. Wilson, E. H. (1913). *op. cit.* Vol. 1. p. 142.
5. Bean, W. J. (1976). *Trees and Shrubs Hardy in the British Isles*. Vol. 3. 8th edn. John Murray. London. p. 317.
6. Archive of the Arnold Arboretum. Wilson Field Journal. Fourth Expedition. 30 August 1910.
7. Wilson, E. H. Unpublished manuscript 'Wilson plants in cultivation'. Archives of the Arnold Arboretum. Series W. XIII. Box 19, Folder 1..
8. Wilson, E. H. Unpublished manuscript. *op. cit.*.
9. Wilson, E. H. (1927). *Plant Hunting*. The Stratford Company. Boston. Massachusetts. Vol. 1 pp. 150–151.
10. Wilson, E. H. (1913). *op. cit.* Vol. 1. p. 113.
11. Wilson, E. H. (1913). *op. cit.*
12. Wilson, E. H. (1913). *op. cit.*

CHAPTER EIGHT: SAVING WILSON'S PLANTS

1. Bean, W J. (1976). *Trees and Shrubs Hardy in the British Isles*. Vol. 1. 8th edn. John Murray. London. p. 15.
2. Lancaster, R. (1991) A Brief History of Plant Hunting. In *The Hillier Manual of Trees and Shrubs*. 6th edn. Hillier Nurseries (Winchester) Ltd. Romsey. Hampshire. p. 33.
3. Rehder, A. (1930). Ernest Henry Wilson. *Journal of the Arnold Arboretum*. 11 (4) 181–192.
4. Shephard, S. (2003). *Seeds of Fortune. A Gardening Dynasty*. Bloomsbury, London. p. 274.
5. Spongberg, S. A. (1990). *A Reunion of Trees*. Harvard University Press. Cambridge. Massachusetts. p. 219.
6. Hay, I. (1995). *Science in the Pleasure Ground. A History of the Arnold Arboretum.* Northeastern University Press. Boston. Massachusetts. p. 166.
7. Hillier, H. G. (1976). *E. H. Wilson*. The Garden. 101 (10) 497–498.
8. Matthews, V. (1993). Wilson's Living Legacy. In Briggs, R W. *'Chinese Wilson'. A Life of Ernest H Wilson 1976–1930*. HMSO. London. pp. 117–130.
9. Bean, W. J. (1976). *Trees and Shrubs Hardy in the British Isles*. Vol. 2. 8th edn. John Murray. London. p. 86.
10. Wilson, E. H. (1927). *Plant Hunting*. The Stratford Company Publishers. Boston Massachusetts. Vol. 2. p. 123.
11. Wilson, E. H. (1926). *Aristocrats of the Garden*. The Stratford Company Publishers. Boston Massachusetts. pp. 221–222.

EPILOGUE

1. 'Quake likely delivered 1-2 punch.' *The Seattle Times* 16 May 2008.
2. http://earthquake.usgs.gov/eqcenter/eqinthenews/2008/us2008ryan
3. 'Beichuan to be laid to rest as China moves survivors to new settlement. *The Times*. 21 May 2008.

SELECTED BIBLIOGRAPHY

Baber, E. C. (1882). Travels and Researches in Western China. *Royal Geographical Society Papers.*

Bean, W. J. (1970–1980). *Trees and Shrubs Hardy in the British Isles.* 8th edition. 4 volumes. John Murray. London.

Bretschneider, E. (1898). *History of European Botanical Discoveries in China.* Sampson Low Marston & Company Ltd. London.

Briggs, R. W. (1993). *'Chinese Wilson'. A Life of Ernest H Wilson 1976–1930.* HMSO. London.

Coats, A. M. (1970). *The Plant Hunters.* McGraw-Hill Book Company. New York.

Cox, E. H. M. (1945). *Plant Hunting in China. A History of the Botanical Exploration of China and the Tibetan Marches.* Collins. London.

Cox, K. ed. (2001). *Frank Kingdon Ward's Riddle of the Tsangpo Gorges.* Antique Collectors' Club. Woodbridge, Suffolk.

Farrington, E. I. (1933). *Ernest H. Wilson Plant Hunter.* The Stratford Company. Boston. Massachusetts.

Fisher, J. (1982). *The Origins of Garden Plants.* Constable. London.

Flanagan, M. and Kirkham, T. (2005). *Plants from the Edge of the World.* Timber Press. Portland, Oregon.

Foley, D. J. (1969). *The Flowering World of 'Chinese' Wilson.* Macmillan. London.

Fox, H. M. (ed). (1949). *Abbé David's Diary.* Harvard University Press. Cambridge. Massachusetts.

Goodman, J. (2006). *Joseph F. Rock and His Shangri-La.* Caravan Press, Hong Kong.

Hay, I. (1995). *Science in the Pleasure Ground: a history of the Arnold Arboretum.* Northeastern University Press. Boston. Massachusetts.

Lancaster, C. R. (1989). *Travels in China: a plantsman's paradise.* Antique Collectors' Club. Woodbridge. Suffolk.

McLean, B. (2004). *George Forrest Plant Hunter.* Antique Collectors' Club. Woodbridge. Suffolk.

Pim, S. (1966). *The Wood and the Trees: a biography of Augustine Henry*. Macdonald and Co. London.

Pratt, A. E. (1892). *To the Snows of Tibet through China*. Longmans. London.

Sargent, C. S. (ed) (1913–1917). *Plantae Wilsonianae*. 3 Volumes. Harvard University Press. Cambridge. Massachusetts.

Schaller, G. B. (1993). *The Last Panda*. The University of Chicago Press. Chicago.

Shipley Cunningham, I. (1984). *Frank N Meyer: plant hunter in Asia*. Iowa State University Press. Ames, Iowa.

Shulman, N. (2002). *A Rage for Rock Gardening*. Short Books. London.

Spongberg, S. A. (1990). *A Reunion of Trees*. Harvard University Press. Cambridge, Massachusetts.

Sutton, S. B. (1970). *Charles Sprague Sargent and the Arnold Arboretum*. Harvard University Press. Cambridge, Massachusetts.

Sutton, S. B. (1974). *In China's Border Provinces: the turbulent career of Joseph Rock, botanist-explorer*. Hastings House. New York.

Whittle, T. (1970). *The Plant Hunters*. William Heinemann Ltd. London.

Wilson, E. H. (1913). *A Naturalist in Western China with Vasculum, Camera and Gun*. 2 Volumes. Methuen and Co Ltd. London.

Wilson, E. H. (1917). *Aristocrats of the Garden*. Doubleday, Page and Co. Garden City, NY.

Wilson, E. H. (1920). *The Romance of Our Trees*. Doubleday, Page and Co. Garden City, NY.

Wilson, E. H. (1927). *Plant Hunting*. 2 Volumes. The Stratford Company. Boston, Massachusetts.

Wilson, E. H. (1928). *More Aristocrats of the Garden*. The Stratford Company. Boston, Massachusetts.

INDEX

Abelia schumannii 215
Abies delavayi 43
Abies fabri **41**, 43
Acer
    A. campestre 213
    A. davidii 238, **241**
    A. griseum **213**
    A. henryi 216
    A. sinense subsp. *wilsonii* 238, **241**
Actinidia chinensis 214
Alnus cremastogyne **48**
An Xian 145, 150
Anlan bridge **140**
Appalachian yellow poplar 218
Arisaema
    A. candidissimum 213
    A. triphyllum 213
Armand, David 23
Arnold Arboretum 10, 18, 22, 24, **25**, 80,
    81, 143, 213, 214, 222
    Journal of the Arnold Arboretum 213
    Keeper of 27
Arnoldia 18, **19**

Bai Cao River 157
Balang Shan 109
Balang Shan Pass 99, 109, **113**, **117**
Batang 67
Bean, W. J. 15, 22, 193, 213, 215
    *Trees and Shrubs Hardy in the British
        Isles*
        15, 19
Beichuan 151, 160, 161, 229, 230, **231**,
    237
    Beichuan Hotel 151, 230, **231**
    bridge 230
Beijing 33, 42, 97, 209
Berberis wilsoniae 9, 238, **241**
Betula albosinensis 10, 216
Big Cannon Mountain(s) 86, **87**
Big Gold River 93
Big Snow Mountains 65
Bond, Ian **24**, **227**
Bond, Caroline **227**

Caltha 116
Carrierea 222
Catalpa
    C. fargesii **23**, 227
    C. ovata **217**
    C. bignonioides 217
Ceratostigma willmottiana 204
Cercidiphyllum japonicum var. *sinense* 160
Chamaecyparis formosensis **16**
Changjiang 42, 43, 129, **130**

Chaoke **34**
Che-shan 150, 151
Chenault, Léon 214, 222
Chengdu 18, 33, 129, 132, 143, 183, 192,
    193, 197, 208, 209
    city wall 209, **210**
    Daoist temple **211**
    Friends Mission 207, **208**
    Guanxianzi **210**, 211
    Institute of Biology 232
    nine arch bridge 209
Chengdu Plain 29, 43, 125, 129, 208
China
    Boxer Rebellion 71
    Confucianism 156
    Cultural Revolution 13, 50, 156, 246
    Emperor Kaiyuan 40
    Emperor Qianlong **97**, 99
        military campaigns 186
    Emperor Qinshihuangdi 135
        Terracotta Army 135
    Empire 34, 154, 208
    ethnic peoples 67
        Ba 134
        Qiang 67, 231
        Khampa 67, **69**
    First Opium War 102
    government 42, 234
    Guomindang troops 50
    Han Chinese 29
    hanyu pinyin 104
    Inland Mission 70, 71
    King Sonom 97
    Macartney embassy 99
    Mao Zhedong 50
        Red Army **41**, 50, 106
        Long March 47, 106
    Ming Dynasty 156
    Ming Emperors 97
    Natural Forest Protection Program
        (NFPP) 42
    Qin 135
    Qing 97
    Qing administration 157
    Qing Dynasty 13
        early 99
        collapse 156
    Sanxingdui, Shu Kingdom 135
    Second Opium War 102
    Sloping Land Conversion Program
        (SLCP) 43
    Sino-Japanese War 187
    Ten Great Campaigns 97
    Treaty of Tientsin 71, 102
Chinese chestnuts **47**

Chinese fir 18, **55**
Chinese flowering dogwood **225**
Chinese hazel 217
Chinese red-barked birch **216**
Chinese tulip tree 218, 219
Chongqing 33, 197, 209
Chu Ba 50
Chungpa 144
Cladrastis
    C. sinensis **218**
    C. kentuckea 218
    C. sinensis 218
    C. wilsonii 218
Clematis montana var. *rubens* 215
common Periwinkle 116
Cornus kousa var. *chinensis* 225
Corydalis 116
Corylopsis 224
    C. sinensis 223, **224**
Corylus chinensis 217
Cotinus coggygria **202**, 204
Cotoneaster dammeri 215
Cox, Kenneth
    Riddle of the Tsangpo Gorge 27
crepe myrtle trees 137
Cunninghamia lanceolata 18, **20**, 55
Cupressus torulosa **90**
Cypripedium
    C. flavum **167**, 172
    C. tibeticum 83, 116

Da Jin Ho 93, 107
Da Pao Shan 86, **87**, 143, 216
Da Xue mountains 28, 69
Da Xue Shan **29**, 65, **71**, 77, 80, 82, 83,
    143
Dadu River 40, **41**, **46**, 47, 50, **51**, 54, 69,
    93, 129
Dadu Valley 47, 55, 56, 143
Dan Yun Xia **238**, **239**
Danba **92**
    stone towers **95**
Darragon, Frederique 95
    'Secret Towers of the Himalayas' 95
    Unicorn Foundation 95
Davidia involucrata **7**, 218
Davidson, Dr 207, 208
Dawei 106, **107**
Delavay's fir 43
Del Tredici, Dr Peter 24
Deutzia 215
Diamond Mountains, Korea 221
Diexi 196, 197
Diexi Lake **196**, 197
dove tree 214, **218**

Dujiangyan 129–132, **134**, **135**, 136, 144, 145, 192, 232
    Flying Crane Hotel **132**, 133, 134
    irrigation system 134–136
    Two Kings Temple (Erwang temple) **129**, **135**, **136**, 137, **139**
Dulverton, Lord 24

earthquake (1933) 196, 197
earthquake (2008) **229**, 229–237
    epicentre 233
    Longmenshan fault 229
    one-way traffic system **233**
East India Company 102
Eastern Han dynasty 132
Edgar, Reverend James Huston 70, 71, 73, 77
Edinburgh, Royal Botanic Garden 15, 22, 226
    Dawyck Botanic Garden 226
    Knott, David, Curator 226
Emei Shan 9, 41, 43
Emmenopterys henryi **221**, 226
Erh-tao chiao 125
Erh-Wang Miao **138**
Erlang Shan 47, 223

Fagaceae 43
Farrer, Reginald 27
Farrington, Edward 27
Feng Zhen Quan 163
field maple 213
Flacourtiaceae 222
Forrest, George 27, 218
Four Girls Mountain 107, **109**
Fu Tang Ba 207
Fubian River **102**
Fubien Ho 106
Fujian River 238

Galearis wardii **118**
Gan Gu River 234
Gansu 129, 185, 217
Ganxipo 45
Giles, Herbert Allen 102
ginkgo **54**
Ginkgo biloba **52**, 54
Gleditsia 133
golden rain tree 204
Gongga Shan **34**, 54, 69, 70, 77, 80, 107
Gorovoy, Peter **22**
Grey-Wilson, Dr Chris 77
Gun Gan Lio Pass 185, **186**

Heathcoat Amory, Michael 24
Henry, Augustine 222, 224
Hien Chu Hsien 145
Hillier, Sir Harold 215
    Hillier Manual of Trees and Shrubs 15, 213, 215

Himalaya, eastern extension of 69
hornbeams 43
Hosie, Sir Alexander 82, 111, 136, 154
    HBM's Consul-General at Chengtu 82
Howick, Charles 216
    Howick Arboretum 10
Hsao Kuan Chai 99, 106
Hsao-chin Ho 100
Hsaochin 104
Hsiang-yang-ping 111, **115**
Hsin Tung Kuan 196
Hsin-tientsze 82
Hu, Mr 35
Hu Jintao 230
Huang He 42, 43
Huanglong: Yellow Dragon Valley 172, 187, **239**
    Dragon Ridge Waterfall 172
    Five-coloured Pond 172
    Mirror-image Pond 172
    Revolving Flower Pond 172
Hubei 14, 70, 214, 221
    western 22, 217, 218

Ichang 30, 221
Idesia 222
Incarvillea
    I. arguta **202**, 204
    I. compacta var. *qinghaensis* 83, **84**
Indian bean tree 217
Iris chrysographes **75**

Je-shui-t'ang 83
Jinjiang River 209
Jinsha River 69
Jiu Yan Qiao **209**
Juglandaceae 24
Juglans 24
    J. regia **106**
Juizhaigou 187, 197
Juniperus tibetica **57**

Kaiping 156, 157, 230, 237
Kangding **18**, 28, **41**, 44, 47, 55, 56, **57**, **61**, **63**, 65, 67, 71, **72**, 80, 81, 82, 130, 185, 187, 218, 222
Kew, Royal Botanic Gardens **15**, 19, **25**, 27, 33, 185, 222, 227
    Berberis Dell 222
    collections 226
    Director of 29
    expeditions 69
    Holly Walk 227
    'T' range 222
    Temperate House 227
    Wakehurst Place 134
Kham 67
Kia-ting Fu 34
Kingdon-Ward, Frank 27, 31
Kiwi fruit 214
Koelreuteria paniculata **202**, 204

Kolkwitzia amabilis **215**
Kuan Hsien 29, 81, 82, 144
Kuei-yung 88
Kung-ching-chiang 145
    village temple **242**
Kwanyin-ping 43

Lagerstroemia indica 137
lampshade poppy 11, 28, 67, 69, 73, **76**, 77, 110, 117
Lancaster, Roy **10**
Larix potaninii 107, **109**, 238, **239**
Lauraceae **132**, 133
Lei-ku-ping 150
Leigu 150
Lengji 47, **51**, 54, 56, 154
Lennox-Boyd, Lady Arabella 24
Leshan 33, 34, 40, 41, 143
    Buddha **40**
Lhasa 33, 67, 70, 71
Li Bing 29, 134–137,140, 232
    revetment system **141**
    Shu State, regional governor 134
Li Er Lang 135
Li Zhongquan 45
Li-tiu Shan 136
Lian hua Shan 72, **73**
Liang, Mr & Mrs **156**, 230, 237
Liang Feng Ya 230
Liang Wei-Dong 156
Liang Xue-Fu 156, 157, 237
Lilium
    L. lophophorum 83, **84**
    L. myriophyllum 143
    L. sargentiae 143
    L. sulphureum 143
    L. regale 143
Liquidambar formosana var. *monticola* **227**
Liriodendron
    L. chinense 218
    L. tulipifera 218, **219**
Lithocarpus 43
Little Gold River 93
Liu, Lao 69
Liu Shu Fen 235, **236**
Loder, Gerald 134, 215
Lonicera
    L. nitida 215
    L. pileata 215
    L. tragophylla 219
Lotus Flower Mountains 72
Luding 47, 56
    iron bridge 41, **46**
Lung-ch'ih-chang 154
Lungan Fu 144, 161
Luo Quan Wan 233, **234**
Lu River 57

Machilus 133
    M. ichangensis 134

Magnolia 215
    M. parviflora 221
    M. sieboldii 221
    M. wilsonii **221**, 222
Malus prattii **227**
Manchu 97
Mandarin 99, 102
Mao Niu 88
Mao Niu River 93
    valley **88**
Maoxian 143, 144, 237
Marco Polo Bridge 209
McNamara, William A. 18, 216, 221
Meconopsis
    M. beamishii 77
    M. grandis 77
    M. henrici 28
    M. henrici var. henrici **31**
    M. integrifolia 11, 28, 67, 77, 116
    M. integrifolia subsp. integrifolia **68**
    M. punicea 11, 28, 185, **186**
    M. racemosa 83, **84**
    M. pseudointegrifolia 77
    M. ×finlayorum (M. integrifolia × M.
        quintuplinervia) 77
Meliosma **161**
    M. beaniana 22, 161, **163**
    M. veitchiorum 22, 222
Mianzhu 145, 244
    outdoor theatre 238, **243**
Micang Shan 154, 216, 217, 221, 222
Mien Chou 144
Min Mountains 129, 144, 161, 166, 172
Min River 29, 33, 40, 47, 69, 94, 125,
        129, 130, 135, 137, 140, 143, 183,
        187, 192, 193, 216
Min Valley 143, 205, **207**, **232**, 234
    lower Min Valley 144, **229**
Miss Willmott's rose 200
Monkong Ting 82, **104**
Mount Chekhov **22**
Mount Everest 70
Mount Omei 9 – see Emei Shan
Moxi 18, 19, 22, 55, 56
Mrs Wilson's Barberry **241**
Myers, Norman 121

Nanjing 102
nanmu 132, 133, 134
    'southern wood' 134
Ni-tou Shan Pass 125
Ning Zhiqi 244

Omei Hsien 154
Omphalogramma vinciflora 117

pagoda 40, 130, **131**, **132**
Pai Shan Yin 197, **199**, 204, 237
'pai-tzu' 183
Pan-lan shan 111, 125

paperbark maple **213**
Parthenocissus henryana **224**
Pedicularis 116
Pekin 67
Piankou 160
Pingwu 144, 160, 161, 238
Pitao River **121**, 124
PlantNetwork of Britain and Ireland 23, 226
Podophyllum hexandrum **75**
Poliothyrsis sinensis 222, **223**
Populus
    P. cathayana 193
    P. suaveolens 193
    P. wilsonii 219
Potanin 185
Pratt, A. E. 154
Primula 117
    P. amethystina 28, **30**
    P. chionantha subsp. sinopurpurea **111**
    P. involucrata subsp. yargonensis **108**
    P. secundiflora 11, **75**
    P. sikkimensis **75**, 116
    P. veitchii 116
    P. vincaeflora 116
Prince Henri d'Orleans **31**
Prunus serrula **213**
Pterocarya 56
    P. hupehensis **56**
Purdom, William 217
Pyracantha atalantioides 215

Qing Yang Gong 211
    temple (Green Goat Temple) 209, 211
Qingcheng Shan 132
Qionglai Shan 107, 109, 183
Quarryhill Botanical Garden (USA) 18,
        216, 221

Red Basin 129
red flag poppy 11, **185**
regal lily **143**, 204, 205, 207, 214
Rehder, Alfred 213, 215, 217
Rheum alexandrae 11, **75**
Rhododendron 215
    R. argyrophyllum 226
    R. balangense 117, **120**
    R. calophytum var. openshawianum 43
    R. capitatum 83
    R. galactinum 117, **120**
    R. orbiculare 219, **220**
    R. oreodoxa 172
    R. przewalskii 11, **81**
    R. rufum 172
    R. schlippenbachii 221
    R. watsonii 172
    R. wiltonii 43
Richards, Shelley 227
Rilong 107, 109
Rock, Joseph 70, 185
Romi Chango 82, 92, 94

Rosa 215
    R. soulieana **201**, 204
    R. willmottiae 200
rowan trees 43, 213
Rubiaceae 221
Rubus xanthocarpus 186
Ruddy, Steve 18

Sakhalin Island **22**
Salix
    S. babylonica 197
    S. magnifica 121
San-chia-tsze 173, 183
San-tsze-yeh 31
Sargent, Charles Sprague 54, 80, 143
Schaller, George 23
Schizophragma 221
Scott, Robert 27
Senecio 116
Shaanxi 33, 50, 156, 237
Shandong Province 237
Shang Tian Ba 50
Shanghai 209
Shanxi 71
Shih Ta Kuan 197
Shih'chuan Hsien 235
Shihch'uan Hsien 144, 150, 151, 154
Shui-ching pu 160
Shuijing 161
Shulman, Nicola 27
Sian Sou Qiao 206, 207
Sichuan 9, **14**, 15, 25
    western 222
    northern 185
    University 207
    Youyang County 221
    Sichuan Basin 229
Sichuan–Tibet highway 45
Siebold, Philippe von 217
'Sifan' (western barbarian) 31, 67
Siguniang Shan 107, **109**
    National Park **108**
    Siguniangshan Hotel 107
Sikkim cowslip 116, 117
Silver Dragon Valley 124, **127**
smoke bush 204
Snow Treasure Peak 144, 238
Soh Chiao 206
Songpan 22, 143, 144, 173, 183,
        185–187, **189**, 192, 193, 197
    covered bridge **192**
    North Gate **187**
Sorbus
    S. americana 213
    S. aucuparia 213
    S. pseudohupehensis 214
    S. setschwanensis 214
Staunton, George 102
Sungpan Ting 144, 186 – see also Songpan
Syringa reflexa 223

Ta Kiang (Great River) 129
Ta-wei 107
Tachienlu **18**, **63**
Tachin Ho 92
Tagong 86
    temple **87**
Taiwan 15, 22, 97, 104
Taning Hsien 151
Tatien-lu 67
Taylor, J. Hudson 70
Tetradium daniellii **227**
Tianjian River 230, 235
Tianjin 209
Tianquan River 47
Tibet 28, 45, 71
    Gyarong 97, 107
Tibetan cherry **213**
Tibetan lady's slipper orchid 83, 117
    Thibetan Lady-slipper orchid 116
Tibetan Plateau **29**
Tibetan uplands 23
Tiled House Mountain 40
Trollius 116
    T. yunnanensis **118**
Tsao shan 41
Tung River 40
Tung-ku 89

Ullung Do (South Korea) 19

Veitch, Harry James 134, 214
Veitch, James Herbert 214
Veitch nursery 67, 77, 80, 143, 214
Viburnum
    V. betulifolium 238, **241**
    V. davidii 215
    V. lantana 225
    V. rhytidophyllum 225
    V. ×rhytidophylloides 225
Vinca major 116
Virginia creeper **224**

Wade, Sir Thomas Francis 102, 154
    Chefoo Convention 154
Wade–Giles system 104
walnut(s) 24, **47**, 106, 197, **200**
Wan-jen-fen 111
Wang Hangming 33, 50, 124, **166**, 183,
        197, 205, 207, 230, 232, 246
Wang Hongbing 246
Wanxian **130**
Wa shan 41, 43
Wawu Shan 40, **41**, 42, 43
Wen Jianbiao 230
Wilson, E. H.
    Aristocrats of the Garden 223
    A Naturalist in Western China 44, 151, 207
    Chinese field journal 206
    Chipping Campden 24, 27

Cotswold house **28**
    family portrait **14**
    horticultural achievements 213
    Plant Hunting 207
    Plantae Wilsonianae 193, 213, 214
    Sanderson camera 50
Wilson's photographs:
    A study in architecture **242**
    Bamboo suspension bridge **148**, **152**
    brick tea **43**
    Friend's Mission Compound **208**
    Hao-chou 197, **198**
    hostel of San-chia-tsza **173**
    Hsao-Ho-Ying **164**
    Hsao-pa-ti **158**
    Hsuan-kou 130, **131**
    Hsueh-po-ting **168**, **174**
    Hsueh-shan Pass **176**, **178**
    Kai-ping-tsen **155**
    Kia-ting Fu **36**, **38**
    Kuan Hsien **112**, **114**, **122**, **133**, **138**
    Lamaseries **64**, 65
    Leng-che **52**
    Lu-ting-chiao **48**
    "many coloured waters" **170**
    Min River **38**
    Min valley South of Sungpan 193, **194**
    "Moorland and crag" **180**
    My Chinese collectors, all faithful and
        true **246**
    Ornate memorial stone **155**
    Pan-lan-shan **112**, **114**
    Pan-lan-shan Valley **122**, 124, **126**
    portrait of his daughter Muriel **144**
    Romi-chango **93**
    Sifan (Sifen) hamlet 197, **198**
    Sungpan Ting **188**, **190**
    Tachien-lu **20**, **43**, **60**, **62**, **78**
    Tachien-lu River **58**
    Taiwan red cypress, Alishan **16**
    Tung River **38**, **48**, **50**
    Ya-chia-k'an **78**
    Yen-Heng **204**
Wilson, Ellen (Nellie) 27
Willmott, Ellen 204
Windsor Great Park 226, 227
    Savill Garden 222

Xi Ling 183
Xi Xiao Long 200, 237
Xia Tian Ba 50
Xian 197
Xiang Fu Miao, temple **145**
    statues of deities **146**
Xiao Ba 157, 161, 230
    greenish slate **160**
Xiao Guan Zhai 99, **102**, 106
Xiaojin **104, 105**
Xiao Jin Ho 93, 97, **99**, **102**, 106, 107

Xiao Jin Ho Valley 99, 104, 136
Xiaoho 166
Xiaojin **104**
Xin Tan 196
    bridge **195**
Xin Xian 55, **56**, **57**
Xinjiang 97
Xizang Autonomous Region 69
Xuebaoding 144, **169**, 172, **182**, 183
Xuebaoshan Jiangzi 238

Ya'an **33**, 40, **41**, 44, 45
    chuan bricks 44
    Tea-Horse Trail 44, 47
    Tea-Horse Trail porters **45**
    statue of tea porter **33**
    Yu Du Hotel 44
Ya-jia Pass 28, **67**, 72, **73**, 77, **79**, **80**
    prayer flags **67**
Yachou Fu 44
Yak Village 88
Yala River 57
Yalong River 69
Yan'an 50
Yangtze River 33, 42, 81, 129, **130**, 143,
        246
Yangtze Valley 129
Yanmen **205**
Yeh-tang **162**
    giant meliosma **162**
Yellow River 42
yellow slipper orchid 172
Yellowstone National Park (USA) 172
Yen-Heng 205
Yibin 33, 129
Yichang 81, 143, 144
Yin Hong 246
Yin Kaipu, Dr 24, 33, 34, **35**, 40, 55, 161,
        183, 196, 230, 232
Yin, Lao 185
Yin Long Gou 124
Yingxiu 233
Yo-tsa 99
Yu Shan (Taiwan) 22
Yue-za 99
Yuli 151, 235, **236**
    Tianjian 'quake lake' **235**
Yunnan 218, 222

Zappey, Walter **124**, 125
Zhang Daoling 132
Zheduo Pass 71, **72**, **82**, 83
    prayer flags **82**
Zheduo River 57
Zhenjiang 102
Zhong Shengxian 33, **35**, 44, 86, 130,
        133, 134, 137, 145, 150, 156, 185,
        206, 209, 230, 232, 234, 237